The Origin of Genetics

A Mendel Source Book

Medal designed by V. A· Kovanič for the Mendel Memorial Celebration of the Czechoslovakian Academy of Sciences, 1965.

The Origin of Genetics

A Mendel Source Book

Edited by CURT STERN and EVA R. SHERWOOD
University of California, Berkeley

W. H. FREEMAN AND COMPANY
San Francisco

Printed in the United States of America.

Library of Congress Catalog Card Number 66-27948
ISBN: 0-7167-0655-5

2 3 4 5 6 7 8 9

Foreword

Gregor Mendel's short treatise "Experiments on Plant Hybrids" is one of the triumphs of the human mind. It does not simply announce the discovery of important facts by new methods of observation and experiment. Rather, in an act of highest creativity, it presents these facts in a conceptual scheme which gives them general meaning. Mendel's paper is not solely a historical document. It remains alive as a supreme example of scientific experimentation and profound penetration of data. It can give pleasure and provide insight to each new reader—and strengthen the exhilaration of being in the company of a great mind at every subsequent study.

Mendel's triumph was a lonely one. Neither his fellow members in the Brünn Natural History Society, nor the readers of its Proceedings or of his reprints were able to understand the significance of his achievement. Not even Carl Nägeli, the famous botanist with whom Mendel corresponded over a period of seven years, grasped the meaning of Mendel's work. Mendel himself, who had followed his experiments with peas by additional ones on hybridization in the genus *Hieracium,* obtained results which did not fit the insight gained by him earlier. Judged by its lack of influence on his contemporaries, his labor and thought seemed a failure. Then, years after his death, the development of biology led to the rediscovery of his findings and interpretations. The failure which had followed his hidden triumph was in turn succeeded by open and lasting vindication.

In his *Critique of Pure Reason,* Immanuel Kant has written as follows: "Observation and analysis of phenomena penetrate into the depths of Nature, and one cannot know how far this will lead in the course of time." The observations and their analysis which Mendel supplied have indeed led far. Without a break in intellectual continuity they became the Mendelism of the early years of this century, joined with the study of chromosomes to expand into the cytogenetics of the next decades, took on new meaning when the biochemical and developmental activities of Mendel's "cell elements" came into the foreground of investigations, and remain conceptually one of the bases of the most advanced contemporary molecular genetics. Mendel's paper of 1866 can be read as a self-contained document illustrating the power of the human intellect to probe into the depths of Nature and it can be read as a prelude to a hundred years of further penetration.

The two genetic papers of Mendel have been available in English translations since 1901 and 1902 respectively. In 1950, on the occasion of the Golden Jubilee of Genetics, English translations of Mendel's letters to Nägeli, as well as of reports of 1900 in which de Vries, Correns, and Tschermak first announced their rediscovery of Mendel's publications on *Pisum* were also published in a supplement to the journal *Genetics.* Now, at the centenary of the experiments on plant hybridization, it seems appropriate to bring together in a single volume the most relevant documents of the period of success, eclipse, and ultimate triumph of Mendelism.

Section I of our collection consists of Mendel's two papers on hybridization and of his letters to Nägeli. In 1965 Professor L. C. Dunn of Columbia University wrote me that Dr. Alan Robertson of the Institute of Animal Genetics, University of Edinburgh, had called his attention to some possible inaccuracies in the original translation of Mendel's main paper commissioned by the Royal Horticultural Society (1901). Professor Dunn checked on some passages and agreed that in at least one instance the translation was inaccurate. In providing an English translation for the present collection of documents, we at first intended to use the translation of 1901 corrected for a few errors. However, a careful comparison with the original German text showed not only a number of mistakes which fundamentally changed the meaning of Mendel's sentences but in addition so many other inaccuracies that Eva Sherwood undertook a completely new translation.

An English translation of Mendel's second paper "On Hieracium Hybrids Obtained by Artificial Fertilisation" (1870) was first made available by Bateson (1902). Here, too, a number of errors (as distinguished from inaccurate shades of rendition) were found. In this case, the original translation has been accepted for printing in the present volume after appropriate correction.

In Mendel's letters to Nägeli he defends his work in objective fashion against Nägeli's speaking of it "with mistrustful caution" (18 April, 1867). He writes that he himself would not have done "otherwise in a similar case" (18 April, 1867). He describes important unpublished new results of genetic significance, among them the proof that a single pollen grain suffices to fertilize an ovule. He also reports to Nägeli on his many crosses between *Hieracium* species, some of which formed the content of his second genetic paper. It is well known that the choice of this genus as a subject for Mendel's work was unfortunate. He could not know—as was shown only many years later—that *Hieracium* is frequently parthenogenetic or apogamous, a situation which obviously results in offspring to which the supposed pollinating parent has not made any genetic contribution.

Mendel's two papers did not sink completely into oblivion. In 1881 the German botanist Focke published a book entitled *Die Pflanzen-Mischlinge*. The bulk of the material was a systematic listing of all the better known plant hybrids.

Under "*Pisum*" and under "*Phaseolus*" Focke cites Mendel's paper of 1866. Similarly, the word "*Hieracium*" in the index leads to a reference to Mendel's 1870 paper.

Focke's specific accounts of the work were all too brief. Moreover, his chapter on the history of hybridization, the last section of which is given to the period from 1851 until "the present" (approximately 1879), is mainly devoted to the work of hybridizers other than Mendel. He refers at length to Naudin and Godron, the two competitors for the 1862 Prize of the French Academy of Sciences on hybridization in the plant kingdom, whose papers were published in the following year. He discusses Wichura's willow hybridizations (1865) and Nägeli's more general treatments of plant hybrids (1865, 1866). In this chapter the only reference to Mendel is contained in a sentence on p. 444: "Among the most recent scientific experiments on hybridization those of Rob. Caspary with *Nymphaeaceae*, of G. Mendel with *Phaseolus* and *Hieracium*, of D. A. Godron with *Datura*, *Aegilops* and *Triticum* and *Papaver* deserve to be called especially instructive." The *Pisum* work is not mentioned here. All of Focke's references to Mendel, translated into English, are presented as our very short Section II.

The simultaneous and independent rediscovery of Mendelism by two investigators, de Vries and Correns, added drama to its long neglect. The communications in which these men announced their findings make up Section III. Actually, de Vries had within a very brief period written three papers which appeared in three different journals. The shortest of these papers, one of two written in French, appeared first in print, but it seems clear that a paper in German was written before the others. It is the German paper that has been translated into English for the present volume. There is no mention of Mendel in de Vries' French note, but his German report refers to Mendel explicitly. Correns was even more emphatic in his recognition of Mendel, as shown by the title of his article: "G. Mendel's Law Concerning the Behavior of Progeny of Varietal Hybrids."

De Vries and Correns had started their own experiments without knowledge of Mendel's publications. It was, however, not surprising that they should find references to his work. Any later worker in this field would naturally consult the more than 38-page-long index of Focke's book at some stage of his inquiries. Since the names of all plants treated in the main body of the work are listed they thus lead to the original references and to Focke's description of any predecessor's findings.

Correns obtained his lead to Mendel's publication directly through Focke's citation in the systematic part of his book. Regarding de Vries, there are two versions of his encountering Mendel's work. In a letter written to Roberts in 1924 de Vries states that he found a reference to Mendel in Bailey's *Plant-Breeding* (1895) "and accordingly looked it up and studied it." Actually, Bailey himself had never seen Mendel's paper but had taken the reference from Focke's writing (Roberts, p. 323). Three decades after de Vries' letter T. J. Stomps (1954), who was de Vries' successor as Professor of Botany at Amsterdam, tells a different story. "In 1900," he writes, "at just the time he [de Vries] was about to publish the results of his experiments he received a letter from his friend, Professor Beyerinck at Delft, reading thus: 'I know that you are studying hybrids, so perhaps the enclosed reprint of 1865 by a certain Mendel which I happen to possess is still of some interest to you.' De Vries read the paper and found that the results of his experiments, which he had believed to be quite new, had already been reported 35 years before. . . . This then is the true story of the rediscovery of Mendel. I once asked de Vries whether he could remember the precise moment at which he discovered Mendel's now famous paper, and he personally related the story to me."

In preparation for his book, *Plant Hybridization before Mendel*, Roberts solicited letters from de Vries and Correns "as to the manner of its [Mendel's paper] discovery by them individually." De Vries answered in English, Correns in German. Both statements were published in English by Roberts and form our Section IV.

The personal accounts of de Vries and Correns stress that they began their experiments before having had any knowledge of Mendel and that they consulted Mendel's paper only after they had come to their own interpretations. Correns' first paper shows clearly his deep understanding of what became known as Mendelism and all his later work bears witness to a mind which could have reached by itself the full gamut of the experimental results and theoretical insights of Mendel. Similarly, de Vries showed a full appreciation of his own and Mendel's findings. This is not surprising, since he was well prepared by his preceding evolutionary studies and his concept of pangenes, which he had advanced much earlier.

It would appear natural to end a collection of Mendeliana at this point. Yet, following a precedent set by Bennett, one more section was added. Its main part is R. A. Fisher's "Has Mendel's Work Been Rediscovered?" (1936). Since the turn of the century Mendel's main paper must have been read by many thousands of people—established investigators and teachers as well as searching students. No one, however, had noticed that Mendel's data were closer to theoretical predictions than would have been expected as a result of chance. This R. A. Fisher found to be true, more than three decades after the rediscovery.

Why Mendel's specific data are "too good" from a statistical point of view remains unknown, but comments which throw some light on this question have kindly been provided by Professor Sewall Wright: these follow the reprint of Fisher's remarkable paper.

This collection of reports and communications does not include material from Tschermak. Although it has been customary to refer to de Vries, Correns, and Tschermak as the three rediscoverers of Mendel, a careful reading of Tschermak's three papers of 1900, each entitled "On artificial hybridisation in *Pisum sativum*," indicates that the designation "rediscoverer" has only limited validity. It is true that all three investigators independently retrieved the forgotten paper on plant hybrids which Gregor Mendel had published. There, however, the common merit ends. De Vries and Correns not only claimed that they had discovered the essential facts and developed their interpretation before they had found Mendel's article but they demonstrated in their own publications that they fully understood the essential aspects of Mendel's theory: the assumption that pairs of determinants were present in hybrids, that they were segregated in germ cells, and that this segregation accounted for the ratios in various types of crosses. On the other hand, Tschermak's papers of 1900 not only lack fundamental analysis of his breeding results but clearly show that he had not developed an interpretation. In an uncertain way and without an attempt at deducing general principles he does present data on F_2 plants indicating 3:1 ratios. When he reports on the results of the back cross of F_1 hybrids to their recessive parents he notes that it "results in an increase of the number of bearers of the recessive trait as compared to the number of recessives obtained after selfing of the F_1." Specifically, he adds that "the influence of the trait 'yellow' in the seeds of the hybrid was reduced by 57 per cent, that of the trait 'smooth' by 43.5 per cent." No discussion of these percentages is offered. The quoted sentences reveal that Tschermak had not realized that back crosses should give a 1:1 ratio.

Twenty-five years later, in his letter to Roberts (published in 1929), Tschermak states that "quite intentionally, I expressed [the rules of inheritance] at first purely descriptively or phenomenologically, in order not at once to anchor the newly-beginning experimental phase of the doctrine of inheritance . . . to definite theoretical terms." Mendel's achievement, however, was that he did go beyond the counting stage and had "indeed entered the rational domain" (Mendel's letter to Nägeli, April 18, 1867; see p. 63 of the present book). Tschermak's influence on the recognition of Mendelian genetics by plant breeders apparently was considerable. Yet his publications in 1900 show him not to have been a rediscoverer of Mendelism but only an experimenter whose understanding—to use a phrase coined by Bateson in another context—had "fallen short of the essential discovery."

In the history of Mendelism the eternal is strongly confounded with the ephemeral. There is the neglect of Mendel's accomplishment during his lifetime and the futility of his years of writing to Nägeli. There is the uneven history of the rediscovery of his work thirty-four years after its publication and its striking reanalysis another thirty-six years later.

It is hoped that collecting in a single book what hitherto has been available only in diverse places will help many new readers to study critically this extraordinary series of events in the history of human thought.

CURT STERN

July, 1966

Acknowledgments

For granting permission to reprint material the editors wish to thank *Genetics* for "Gregor Mendel's Letters to Carl Nägeli," and for C. G. Correns' "Mendel's Law Concerning the Behavior of Progeny of Varietal Hybrids," both from Birth of Genetics (*Genetics* 35, Suppl.): 1–29, 33–41, 1950; Taylor & Francis Ltd., London, for R. A. Fisher's "Has Mendel's Work Been Rediscovered?" *Annals of Science* 1:115–137, 1936; and Princeton University Press for excerpts from H. F. Roberts' *Plant Hybridization before Mendel,* pp. 323, 335–338, Princeton University Press, Copyright 1929.

The photograph of Hugo de Vries was supplied by Prof. T. J. Stomps, University of Amsterdam, the one of Carl Correns by Dr. A. E. Correns, Munich, and the one of Ronald A. Fisher by Dr. W. J. Youden, Washington, D. C. Their help has made it possible to include likenesses of authors represented in the present volume taken at the time of their relevant contribution.

We are grateful to our colleagues and friends for their valuable discussions and suggestions during the preparation of this book.

Contents

The Origin of Genetics

A Mendel Source Book

Part One 1866-1873

Experiments on Plant Hybrids

Versuche über Pflanzen-Hybriden

GREGOR MENDEL

Read at the Meetings of 8 February and 8 March, 1865

Translated by Eva R. Sherwood[1]

INTRODUCTORY REMARKS

Artificial fertilization undertaken on ornamental plants to obtain new color variants initiated the experiments to be discussed here. The striking regularity with which the same hybrid forms always reappeared whenever fertilization between like species took place suggested further experiments whose task it was to follow the development of hybrids in their progeny.

Numerous careful observers, such as Kölreuter, Gärtner, Herbert, Lecoq, Wichura, and others, have devoted a part

[The original paper was published in Verhandlungen des naturforschenden Vereines in Brünn 4 (1865), Abhandlungen, pp. 3–47, which appeared in 1866.]

[1] [An attempt has been made to preserve Mendel's style; frequently however, it was necessary to add a word or change the order of an English sentence to permit the reader insight into the clarity with which the German version was written. The abbreviations used correspond to those in the printed publication; they are not always identical with the ones in the handwritten manuscript. A historian in genetics

of their lives to this problem with tireless persistence. Gärtner, especially, in his work "Die Bastarderzeugung im Pflanzenreich" (Hybrid Production in the Plant Kingdom) has recorded very estimable observations, and Wichura has very recently published the results of his thorough investigations of willow hybrids. That no generally applicable law of the formation and development of hybrids has yet been successfully formulated can hardly astonish anyone who is acquainted with the extent of the task and who can appreciate the difficulties with which experiments of this kind have to contend. A final decision can be reached only when the results of detailed experiments from the most diverse plant families are available. Whoever surveys the work in this field will come to the conviction that among the numerous experiments not one has been carried out to an extent or in a manner that would make it possible to determine the number of different forms in which hybrid progeny appear, permit classification of these forms in each generation with certainty, and ascertain their numerical interrelationships. It requires a good deal of courage indeed to undertake such a far-reaching task; however, this seems to be the one correct way of finally reaching the solution to a question whose significance for the evolutionary history of organic forms must not be underestimated.

This paper discusses the attempt at such a detailed experiment. It was expedient to limit the experiment to a fairly small group of plants, and after a period of eight years it is now essentially concluded. Whether the plan by which the

---

would, of course, want to consult the original paper.

Throughout this translation "differierend" has been translated by "differing," "Merkmal" by "trait," "Character" by "character" or "characteristic," "Entwicklungsreihe" by "series," "je zwei" by "one pair," "dominierend" by "dominating," and "Glied" by "term" when used in an algebraic series, by "member" when the word applies to a series of genotypes.

While this terminology needs no further comment, some explanation of the use of "angular" for "kantig" is necessary. In 1901 the Journal of the Royal Horticultural Society of London published a trans-individual experiments were set up and carried out was adequate to the assigned task should be decided by a benevolent judgment.

SELECTION OF EXPERIMENTAL PLANTS

The value and validity of any experiment are determined by the suitability of the means used as well as by the way they are applied. In the present case as well, it can not be unimportant which plant species were chosen for the experiments and how these were carried out.

Selection of the plant group for experiments of this kind must be made with the greatest possible care if one does not want to jeopardize all possibility of success from the very outset.

The experimental plants must necessarily

1. Possess constant differing traits.

2. Their hybrids must be protected from the influence of all foreign pollen during the flowering period or easily lend themselves to such protection.

3. There should be no marked disturbances in the fertility of the hybrids and their offspring in successive generations.

Contamination with foreign pollen that might take place during the experiment without being recognized would lead to quite erroneous conclusions. Occasional forms with reduced fertility or complete sterility, which occur among the offspring of many hybrids, would render the experiments

---

lation of Mendel's publication, in which the word "wrinkled" describes the shape of seeds whenever "kantig" is the only adjective. English scientific literature has generally followed this precedent and referred to round vs. wrinkled peas. But Mendel describes the seeds of *P. quadratum* as both angular and wrinkled, using two different German words. Classifying the seeds of hybrid plants he uses only the word for "angular" to contrast with that for "round." A search through English botanical texts confirms the accuracy of this description: the seeds are irregularly shaped, asymmetrically compressed, with smooth surfaces meeting at an angle. *E.R.S.*]

very difficult or defeat them entirely. To discover the relationships of hybrid forms to each other and to their parental types it seems necessary to observe *without exception all* members of the series of offspring in each generation.

From the start, special attention was given to the *Leguminosae* because of their particular floral structure. Experiments with several members of this family led to the conclusion that the genus *Pisum* had the qualifications demanded to a sufficient degree. Some quite distinct forms of this genus possess constant traits that are easily and reliably distinguishable, and yield perfectly fertile hybrid offspring from reciprocal crosses. Furthermore, interference by foreign pollen cannot easily occur, since the fertilizing organs are closely surrounded by the keel, and the anthers burst within the bud; thus the stigma is covered with pollen even before the flower opens. This fact is of particular importance. The ease with which this plant can be cultivated in open ground and in pots, as well as its relatively short growth period, are further advantages worth mentioning. Artificial fertilization is somewhat cumbersome, but it nearly always succeeds. For this purpose the not yet fully developed bud is opened, the keel is removed, and each stamen is carefully extracted with forceps, after which the stigma can be dusted at once with foreign pollen.

From several seed dealers a total of 34 more or less distinct varieties of peas were procured and subjected to two years of testing. In one variety a few markedly deviating forms were noticed among a fairly large number of like plants. These, however, did not vary in the following year and were exactly like another variety obtained from the same seed dealer; no doubt the seeds had been accidentally mixed. All other varieties yielded quite similar and constant offspring; at least during the two test years no essential change could be noticed. Twenty-two of these varieties were selected for fertilization and planted annually throughout the entire experimental period. They remained stable without exception.

Their systematic classification is difficult and uncertain. If one wanted to use the strictest definition of species, by which only those individuals that display identical traits under identical conditions belong to a species, then no two could be counted as one and the same species. In the opinion of experts, however, the majority belong to the species *Pisum sativum;* while the remaining ones were regarded and described either as sub-species of *P. sativum,* or as separate species, such as *P. quadratum, P. saccharatum,* and *P. umbellatum.* In any event, the rank assigned to them in a classification system is completely immaterial to the experiments in question. Just as it is impossible to draw a sharp line between species and varieties, it has been equally impossible so far to establish a fundamental difference between the hybrids of species and those of varieties.

ARRANGEMENT AND SEQUENCE OF EXPERIMENTS

When two plants, constantly different in one or several traits, are crossed, the traits they have in common are transmitted unchanged to the hybrids and their progeny, as numerous experiments have proven; a pair of differing traits, on the other hand, are united in the hybrid to form a new trait, which usually is subject to changes in the hybrid's progeny. It was the purpose of the experiment to observe these changes for each pair of differing traits, and to deduce the law according to which they appear in successive generations. Thus the study breaks up into just as many separate experiments as there are constantly differing traits in the experimental plants.

The various forms of peas selected for crosses showed differences in length and color of stem; in size and shape of leaves; in position, color, and size of flowers; in length of flower stalks; in color, shape, and size of pods; in shape and size of seeds; and in coloration of seed coats and albumen. However, some of the traits listed do not permit a definite and sharp separation, since the difference rests on a "more or less" which is often difficult to define. Such traits were not usable for individual experiments; these had to be limited to characteristics which stand out clearly and decisively in the plants. The result should ultimately show whether in hybrid unions the traits all observe concordant behavior, and whether one can also make a decision about those traits which have minor significance in a classification.

The traits selected for experiments relate:

1. *To the difference in the shape of the ripe seeds.* These are either round or nearly round, with depressions, if any occur on the surface, always very shallow; or they are irregularly angular and deeply wrinkled (*P. quadratum*).

2. *To the difference in coloration of seed albumen* (endosperm).[2] The albumen of ripe seeds is either pale yellow, bright yellow and orange, or has a more or less intense green color. This color difference is easily recognizable in the seeds because their coats are transparent.

3. *To the difference in coloration of the seed coat.* This is either white, in which case it is always associated with white flower color; or it is grey, grey-brown, leather-brown with or without violet spotting, in which case the color of the standard is violet, that of the wings is purple, and the stem bears reddish markings at the leaf axils. The grey seed coats turn black-brown in boiling water.

4. *To the difference in shape of the ripe pod.* This is either smoothly arched and never constricted anywhere, or deeply constricted between the seeds and more or less wrinkled (*P. saccharatum*).

5. *To the difference in color of the unripe pod.* It is either colored light to dark green or vivid yellow, which is also the coloration of stalks, leaf-veins, and calyx.[3]

---

[2] ["Mendel uses the terms 'albumen' and 'endosperm' somewhat loosely to denote the cotyledons, containing food-material, within the seed." W. *Bateson.*]

[3] One variety has a beautiful brownish-red pod color which tends to

6. *To the difference in position of flowers.* They are either axillary, that is, distributed along the main stem, or they are terminal, bunched at the end of the stem and arranged in what is almost a short cyme; if the latter, the upper part of the stem is more or less enlarged in cross section (*P. umbellatum*).

7. *To the difference in stem length.* The length of the stem varies greatly in individual varieties; it is, however, a constant trait for each, since in healthy plants grown in the same soil it is subject to only insignificant variations. In experiments with this trait, the long stem of 6 to 7′ was always crossed with the short one of ¾′ to 1½′ to make clear-cut distinction possible.[4]

Each of the two differing traits listed above as pairs were united by fertilization.

For the

1st	experiment	60	fertilizations	on	15	plants	were undertaken.
2nd	"	58	"	"	10	"	" "
3rd	"	35	"	"	10	"	" "
4th	"	40	"	"	10	"	" "
5th	"	23	"	"	5	"	" "
6th	"	34	"	"	10	"	" "
7th	"	37	"	"	10	"	" "

From a fairly large number of plants of the same kind only the most vigorous were chosen for fertilization. Weak plants always give uncertain results, because many of the offspring either fail to flower entirely or form only few and inferior seeds even in the first generation of hybrids, and still more do so in the following one.

---

a violet and blue around the time of ripening. The experiment with this trait was started only in the past year.

[4] [Presumably the symbol ′ stands for Viennese foot, equal to 0.316 meter, or 12.44 inches. *E.R.S.*]

Furthermore, in all experiments reciprocal crosses were made in such a manner that that one of the two varieties serving as seed plant in one group of fertilizations was used as pollen plant in the other group.

The plants were grown in garden beds—except for a few in pots—and were maintained in their natural upright position by means of sticks, twigs, and taut strings. For each experiment a number of the potted plants were placed in a greenhouse during the flowering period; they were to serve as controls for the main experiment in the garden against possible disturbance by insects. Among the insects that visit pea plants the beetle *Bruchus pisi* might become dangerous to the experiment should it appear in fairly large numbers. It is well known that the female of this species lays her eggs in the flower and thereby opens the keel; on the tarsi of one specimen caught in a flower some pollen cells could clearly be seen under the hand lens. Mention must also be made here of another circumstance that might possibly lead to the admixture of foreign pollen. For in some rare cases it happens that certain parts of an otherwise quite normally developed flower are stunted, leading to partial exposure of the fertilization organs. Thus, defective development of the keel, which left pistil and anthers partly uncovered, was observed. It also sometimes happens that the pollen does not fully mature. In that event the pistil gradually lengthens during the flowering period until the stigma protrudes from the tip of the keel. This curious phenomenon has also been observed in hybrids of *Phaseolus* and *Lathyrus*.

The risk of adulteration by foreign pollen is, however, a very slight one in *Pisum*, and can have no influence whatsoever on the overall result. Among more than 10,000 carefully examined plants there were only a very few in which admixture had doubtlessly occurred. Since such interference was never noticed in the greenhouse, it may be assumed that *Bruchus pisi*, and perhaps also the cited abnormalities in floral structure, are to blame.

THE FORM OF THE HYBRIDS

Experiments on ornamental plants undertaken in previous years had proven that, as a rule, hybrids do not represent the form exactly intermediate between the parental strains. Although the intermediate form of some of the more striking traits, such as those relating to shape and size of leaves, pubescence of individual parts, and so forth, is indeed nearly always seen, in other cases one of the two parental traits is so preponderant that it is difficult, or quite impossible, to detect the other in the hybrid.

The same is true for *Pisum* hybrids. Each of the seven hybrid traits either resembles so closely one of the two parental traits that the other escapes detection, or is so similar to it that no certain distinction can be made. This is of great importance to the definition and classification of the forms in which the offspring of hybrids appear. In the following discussion those traits that pass into hybrid association entirely or almost entirely unchanged, thus themselves representing the traits of the hybrid, are termed *dominating*, and those that become latent in the association, *recessive*. The word "recessive" was chosen because the traits so designated recede or disappear entirely in the hybrids, but reappear unchanged in their progeny, as will be demonstrated later.

All experiments proved further that it is entirely immaterial whether the dominating trait belongs to the seed or pollen plant; the form of the hybrid is identical in both cases. This interesting phenomenon was also emphasized by Gärtner, with the remark that even the most practiced expert is unable to determine from a hybrid which of the two species crossed was the seed plant and which the pollen plant.

Of the differing traits utilized in the experiments the following are dominating:

1. The round or nearly round seed shape with or without shallow depressions.

2. The yellow coloration of seed albumen [cotyledons].

3. The grey, grey-brown, or leather-brown color of the seed coat, associated with violet-red blossoms and reddish spots in the leaf axils.

4. The smoothly arched pod shape.

5. The green coloration of the unripe pod, associated with the same color of stem, leaf veins, and calyx.

6. The distribution of flowers along the stem.

7. The length of the longer stem.

With respect to this last trait it must be noted that the stem of the hybrid is usually longer than the longer of the two parental stems, a fact which is possibly due only to the great luxuriance that develops in all plant parts when stems of very different lengths are crossed. Thus, for instance, in repeated experiments, hybrid combinations of stems 1′ and 6′ long yielded, without exception, stems varying in length from 6′ to 7½′. *Hybrid seed coats* are often more spotted; the spots sometimes coalesce into rather small bluish-purple patches. Spotting frequently appears even when it is absent as a parental trait.

The hybrid forms of *seed shape* and *albumen* develop immediately after artificial fertilization merely through the influence of the foreign pollen. Therefore they can be observed in the first year of experimentation, while the remaining traits do not appear in the plants raised from fertilized seeds until the following year.

THE FIRST GENERATION FROM HYBRIDS

In this generation, *along with the dominating* traits, the *recessive* ones also reappear, their individuality fully revealed, and they do so in the decisively expressed average proportion of 3:1, so that among each four plants of this generation three receive the dominating and one the recessive characteristic. This is true, without exception, of all traits included in the experiments. The angular, wrinkled seed shape, the green coloration of the albumen, the white color of seed coat and flower, the constrictions on the pods, the yellow color of the immature pod, stalk, calyx, and leaf veins, the almost umbellate inflorescence, and the dwarfed stem all reappear in the numerical proportion given, without any essential deviation. *Transitional forms were not observed in any experiment.*

Since the hybrids resulting from reciprocal crosses were of identical appearance and showed no noteworthy deviation in their subsequent development, the results from both crosses may be totaled in each experiment. The numerical proportions obtained for each pair of differing traits are as follows:

Experiment 1. Seed shape. From 253 hybrids 7324 seeds were obtained in the second experimental year. Of them, 5474 were round or roundish and 1850 angular wrinkled. This gives the ratio 2.96:1.

Experiment 2. Albumen coloration. 258 plants yielded 8023 seeds, 6022 yellow and 2001 green; their ratio, therefore, is 3.01:1.

In these two experiments each pod usually yielded both kinds of seed. In well-developed pods that contained, on the average, six to nine seeds, all seeds were fairly often round (Experiment 1) or all yellow (Experiment 2); on the other hand, no more than 5 angular or 5 green ones were ever observed in one pod. It seems to make no difference whether the pods develop earlier or later in the hybrid, or whether they grow on the main stem, or on an axillary one. In the pods first formed by a small number of plants only a few seeds developed, and these possessed only one of the two traits; in the pods developing later, however, the proportion remained normal. The distribution of traits also varies in individual plants, just as in individual pods. The first ten members of both series of experiments may serve as an illustration:

	Experiment 1 Shape of Seeds		*Experiment 2* Coloration of Albumen	
Plant	Round	Angular	Yellow	Green
1	45	12	25	11
2	27	8	32	7
3	24	7	14	5
4	19	10	70	27
5	32	11	24	13
6	26	6	20	6
7	88	24	32	13
8	22	10	44	9
9	28	6	50	14
10	25	7	44	18

Extremes observed in the distribution of the two seed traits in a *single* plant were, in Experiment 1, one instance of 43 round and only 2 angular, another of 14 round and 15 angular seeds. In Experiment 2 there was found an instance of 32 yellow and only 1 green, but also one of 20 yellow and 19 green seeds.

These two experiments are important for the determination of mean ratios, which make it possible to obtain very meaningful averages from a fairly small number of experimental plants. However, in counting the seeds, especially in Experiment 2, some attention is necessary, since in individual seeds of some plants the green coloration of the albumen is less developed, and can be easily overlooked at first. The cause of partial disappearance of the green coloration has no connection with the hybrid character of the plants, since it occurs also in the parental plant; furthermore, this peculiarity is restricted to the individual and not inherited by the offspring. In luxuriant plants this phenomenon was noted quite frequently. Seeds damaged during their development by insects often vary in color and shape; with a little practice in sorting, however, mistakes are easy to avoid. It is almost superfluous to mention that the pods must remain on the plants until they are completely ripe and dry, for only then are the shape and color of the seeds fully developed.

Experiment 3. Color of seed coat. Among 929 plants 705 bore violet-red flowers and grey-brown seed coats; 224 had white flowers and white seed coats; this yields the proportion 3.15:1.

Experiment 4. Shape of pods. Of 1181 plants 882 had smoothly arched pods, 299 constricted ones. Hence the ratio 2.95:1.

Experiment 5. Coloration of unripe pods. The experimental plants numbered 580, of which 428 had green and 152 yellow pods. Consequently, the former stand to the latter in the proportion 2.82:1.

Experiment 6. Position of flowers. Among 858 cases, 651 had axillary flowers and 207 terminal ones. Consequently, the ratio is 3.14:1.

Experiment 7. Length of stem. Of 1064 plants, 787 had the long stems, 277 the short stems. Hence a relative proportion of 2.84:1. In this experiment the dwarfed plants were tenderly lifted and transferred to beds of their own. This precaution was necessary because they would have become stunted growing amidst their tall brothers and sisters. Even in the earliest stages they can be distinguished by their compact growth and thick dark-green leaves.

When the results of all experiments are summarized, the average ratio between the number of forms with the dominating trait and those with the recessive one is 2.98:1, or 3:1.

The dominating trait can have *double significance* here—namely that of the parental characteristic or that of the hybrid trait. In which of the two meanings it appears in each individual case only the following generation can decide. As parental trait it would pass unchanged to all of the offspring; as hybrid trait, on the other hand, it would exhibit the same behavior as it did in the first generation.

Those forms that receive the recessive character in the first generation do not vary further in the second with respect to this trait; they remain *constant* in their progeny.

The situation is different for those possessing the dominating trait in the first generation. Of these, *two* parts yield offspring that carry the dominating and the recessive trait in the proportion of 3:1, thus showing exactly the same behavior as the hybrid forms; only *one* part remains constant for the dominating trait.

The individual experiments yielded results as follows:

Experiment 1. Among 565 plants raised from round seeds of the first generation, 193 yielded only round seeds, and therefore remained constant in this trait; 372, however, produced both round and angular seeds in the proportion of 3:1. Therefore, the number of hybrids compared to that of the constant breeding forms is as 1.93:1.

Experiment 2. Of 519 plants raised from seeds whose albumen had yellow coloration in the first generation, 166 yielded exclusively yellow, while 353 yielded yellow and green seeds in the proportion 3:1. Therefore, a partition into hybrid and constant forms resulted in the proportion 2.13:1.

For each of the subsequent experiments, 100 plants that possessed the dominating trait in the first generation were selected, and in order to test this trait's significance 10 seeds from each plant were sown.

Experiment 3. The offspring of 36 plants yielded exclusively grey-brown seed coats; from 64 plants came some grey-brown and some white coats.

Experiment 4. The offspring of 29 plants had only smoothly arched pods; of 71, on the other hand, some had smoothly arched and some constricted ones.

Experiment 5. The offspring of 40 plants had only green pods; those of 60 plants had some green, some yellow.

Experiment 6. The offspring of 33 plants had only axillary flowers; of another 67, on the other hand, some had axillary, some terminal flowers.

Experiment 7. The offspring of 28 plants received the long stem, those of 72 plants partly the long, partly the short one.

In each of these experiments a certain number of plants with the dominating trait become constant. In evaluating the proportion in which forms with the constantly persisting trait segregate, the first two experiments are of special importance, since in these a fairly large number of plants could be compared. The ratios 1.93:1 and 2.13:1 taken together give almost exactly the average ratio of 2:1. Experiment 6 gives an entirely concordant result; in other experiments the proportion varies more or less, as was to be expected with the small number of 100 experimental plants. Experiment 5, which showed the greatest deviation, was repeated, and, instead of 60:40, the ratio of 65:35 was obtained. *Accordingly, the average ratio of 2:1 seems ensured.* It is thus proven that of those forms which possess the dominating trait in the first generation, two parts carry the hybrid trait, but the one part with the dominating trait remains constant.

The ratio of 3:1 in which the distribution of the dominating and recessive traits takes place in the first generation therefore resolves itself into the ratio of 2:1:1 in all experiments if one differentiates between the meaning of the dominating trait as a hybrid trait and as a parental character. Since the members of the first generation originate directly from the seeds of the hybrids, *it now becomes apparent that of the seeds formed by the hybrids with one pair of differing traits, one half again develop the hybrid form while the other half yield plants that remain constant and receive the dominating and the recessive character in equal shares.*

THE SUBSEQUENT GENERATIONS FROM HYBRIDS

The proportions in which the descendants of hybrids develop and split up in the first and second generations are probably valid for all further progeny. Experiments 1 and 2 have by now been carried through six generations, 3 and

7 through five, and 4, 5 and 6 through four without any deviation becoming apparent, although from the third generation on a small number of plants were used. In each generation the offspring of the hybrids split up into hybrid and constant forms according to the ratios 2:1:1.

If A denotes one of the two constant traits, for example, the dominating one, *a* the recessive, and A*a* the hybrid form in which both are united, then the expression

$$A + 2Aa + a$$

gives the series for the progeny of plants hybrid in a pair of differing traits.

The observation made by Gärtner, Kölreuter, and others, that hybrids have a tendency to revert to the parental forms, is also confirmed by the experiments discussed. It can be shown that the numbers of hybrids derived from one fertilization decrease significantly from generation to generation as compared to the number of newly constant forms and their progeny, yet they can never disappear entirely. If one assumes, on the average, equal fertility for all plants in all generations, and if one considers, furthermore, that half of the seeds that each hybrid produces yield hybrids again while in the other half the two traits become constant in equal proportions, then the numerical relationships for the progeny in each generation follow from the tabulation below, where A and *a* again denote the two parental traits and A*a* the hybrid form. For brevity's sake one may assume that in each generation each plant supplied only four seeds.

				Expressed in terms of ratios
Generation	A	A*a*	*a*	A : A*a* : *a*
1	1	2	1	1 : 2 : 1
2	6	4	6	3 : 2 : 3
3	28	8	28	7 : 2 : 7
4	120	16	120	15 : 2 : 15
5	496	32	496	31 : 2 : 31
n				$2^n - 1 : 2 : 2^n - 1$

In the tenth generation, for example, $2^n - 1 = 1023$. Therefore, of each 2048 plants arising in this generation, there are 1023 with the constant dominating trait, 1023 with the recessive one, and only 2 hybrids.

THE OFFSPRING OF HYBRIDS IN WHICH SEVERAL DIFFERING TRAITS ARE ASSOCIATED

In the experiments discussed above, plants were used which differed in only one essential trait. The next task consisted in investigating whether the law of development thus found would also apply to a pair of differing traits when several different characteristics are united in the hybrid through fertilization.

The experiments demonstrated throughout that in such a case the hybrids always resemble more closely that one of the two parental plants which possesses the greater number of dominating traits. If, for instance, the seed plant has a short stem, terminal white flowers, and smoothly arched pods, and the pollen plant has a long stem, violet-red lateral flowers, and constricted pods, then the hybrid reminds one of the seed plant only in pod shape, and the remaining traits resemble those of the pollen plant. Should one of the two parental types possess only dominating traits, then the hybrid is hardly or not at all distinguishable from it.

Two experiments were carried out with a larger number of plants. In the first experiment the parental plants differed in seed shape and coloration of albumen; in the second in seed shape, coloration of albumen, and color of seed coat. Experiments with seed traits lead most easily and assuredly to success.

To simplify a survey of the data, the differing traits of the seed plant will be indicated in these experiments by A, B, C, those of the pollen plant by *a, b, c,* and the hybrid forms of these traits by *Aa, Bb* and *Cc.*

First Experiment:

AB seed plant,	*ab* pollen plant,
A shape round,	*a* shape angular,
B albumen yellow,	*b* albumen green.

The fertilized seeds were round and yellow, resembling those of the seed plant. The plants raised from them yielded seeds of four kinds, frequently lying together in one pod. From 15 plants a total of 556 seeds were obtained, and of these there were:

315 round and yellow,
101 angular and yellow,
108 round and green,
32 angular and green.

All were planted in the following year. Eleven of the round yellow seeds did not germinate and three plants from such seeds did not attain fruition. Of the remaining plants:

38	had round yellow seeds	*AB*
65	had round yellow and green seeds . .	*ABb*
60	had round yellow and angular yellow seeds	*AaB*
138	had round yellow and green and angular yellow and green seeds	*AaBb*

Of plants grown from angular yellow seeds, 96 bore fruit; of these

28	had only angular yellow seeds . . .	*aB*
68	had only angular yellow and green seeds	*aBb*

Of plants grown from 108 round green seeds, 102 bore fruit; of these

35 had only round green seeds . . . . *Ab*
67 had round and angular green seeds . . *Aab*

The angular green seeds yielded 30 plants with identical seeds throughout; they remained constant . . *ab*

Thus the offspring of hybrids appear in nine different forms, some of them in very unequal numbers. When these are summarized and arranged in order, one obtains:

38	plants	with	the designation	*AB*
35	"	"	"	*Ab*
28	"	"	"	*aB*
30	"	"	"	*ab*
65	"	"	"	*ABb*
68	"	"	"	*aBb*
60	"	"	"	*AaB*
67	"	"	"	*Aab*
138	"	"	"	*AaBb*

All forms can be classified into three essentially different groups. The first comprises those with designations *AB*, *Ab*, *aB*, *ab*; they possess only constant traits and do not change any more in following generations. Each one of these forms is represented 33 times on the average. The second group contains the forms *ABb*, *aBb*, *AaB*, *Aab*; these are constant for one trait, hybrid for the other, and in the next generation vary only with respect to the hybrid trait. Each of them appears 65 times on the average. The form *AaBb* occurs 138 times, is hybrid for both traits, and behaves exactly like the hybrid from which it is descended.

Comparing the numbers in which the forms of these groups occur, one cannot fail to recognize the average proportions of 1:2:4. The numbers 33, 65, 138 give quite satisfactory approximations to the numerical proportions 33, 66, 132.

Accordingly, the series consists of nine terms. Four of them occur once each and are constant for both traits; the forms *AB*, *ab* resemble the parental types, the other two represent the other possible constant combinations between the associated traits A, *a*, *B*, *b*. Four terms occur twice each and are constant for one trait, hybrid for the other. One term appears four times and is hybrid for both traits. When, therefore, two kinds of differing traits are combined in hybrids, the progeny develop according to the expression:

$$AB + Ab + aB + ab + 2ABb + 2aBb + 2AaB + 2Aab + 4AaBb.$$

Indisputably this series is a combination series in which the two series for the traits A and *a*, *B* and *b* are combined term by term. All the terms of the series are obtained through a combination of the expressions:

$$A + 2Aa + a$$
$$B + 2Bb + b$$

Second Experiment:

ABC seed plant,	*abc* pollen plant.
A shape round,	*a* shape angular.
B albumen yellow,	*b* albumen green.
C seed coat grey-brown,	*c* seed coat white.

This experiment was conducted in a manner quite similar to that used in the preceding one. Of all experiments it required the most time and effort. From 24 hybrids a total of 687 seeds was obtained, all of which were spotted, colored grey-brown or grey-green, and round or angular. Of the plants grown from them, 639 bore fruit in the following year and, as further investigations showed, they comprised:

8	plants	*ABC*	22	plants	*ABCc*	45	plants	*ABbCc*
14	"	*ABc*	17	"	*AbCc*	36	"	*aBbCc*
9	"	*AbC*	25	"	*aBCc*	38	"	*AaBCc*
11	"	*Abc*	20	"	*abCc*	40	"	*AabCc*
8	"	*aBC*	15	"	*ABbC*	49	"	*AaBbC*
10	"	*aBc*	18	"	*ABbc*	48	"	*AaBbc*
10	"	*abC*	19	"	*aBbC*			
7	"	*abc*	24	"	*aBbc*			
			14	"	*AaBC*	78	"	*AaBbCc*
			18	"	*AaBc*			
			20	"	*AabC*			
			16	"	*Aabc*			

The series comprises 27 members. Of these 8 are constant for all traits, and each occurs 10 times on the average; 12 are constant for two traits, hybrid for the third; each appears 19 times on the average; 6 are constant for one trait, hybrid for the other two; each of these turns up 43 times on the average; one form occurs 78 times and is hybrid for all traits. The ratios 10:19:43:78 approach the ratios 10:20:40:80, or 1:2:4:8, so closely that the latter doubtlessly represent the correct values.

The development of hybrids whose parents differ in three traits thus takes place in accord with the expression:

ABC + *ABc* + *AbC* + *Abc* + *aBC* + *aBc* + *abC* + *abc* + 2*ABCc* + 2*AbCc* + 2*aBCc* + 2*abCc* + 2*ABbC* + 2*ABbc* + 2*aBbC* + 2*aBbc* + 2*AaBC* + 2*AaBc* + 2*AabC* + 2*Aabc* + 4*ABbCc* + 4*aBbCc* + 4*AaBCc* + 4*AabCc* + 4*AaBbC* + 4*AaBbc* + 8*AaBbCc*.

Here, too, is a combination series in which the series for traits A and *a*, B and *b*, C and *c* are combined with each other. The expressions:

A + 2A*a* + *a*
B + 2B*b* + *b*
C + 2C*c* + *c*

supply all terms of the series. The constant associations encountered in it correspond to all possible combinations of the traits A, *B*, *C*, *a*, *b*, *c*; two of them, *ABC* and *abc*, resemble the two parental plants.

In addition, several more experiments were carried out with a smaller number of experimental plants in which the remaining traits were combined by twos or threes in hybrid fashion; all gave approximately equal results. Therefore there can be no doubt that for all traits included in the experiment this statement is valid: *The progeny of hybrids in which several essentially different traits are united represent the terms of a combination series in which the series for each pair of differing traits are combined.* This also shows at the same time that *the behavior of each pair of differing traits in a hybrid association is independent of all other differences in the two parental plants.*

If n designates the number of characteristic differences in the two parental plants, then 3^n is the number of terms in the combination series, 4^n the number of individuals that belong to the series, and 2^n the number of combinations that remain constant. For instance, when the parental types differ in four traits the series contains $3^4 = 81$ terms, $4^4 = 256$ individuals, and $2^4 = 16$ constant forms; stated differently, among each 256 offspring of hybrids there are 81 different combinations, 16 of which are constant.

All constant associations possible in *Pisum* through combition of the above-mentioned seven characteristic traits were actually obtained through repeated crossing. Their number is given by $2^7 = 128$. At the same time this furnishes factual proof that *constant traits occurring in different forms of a plant kindred can, by means of repeated artificial fertilization, enter into all the associations possible within the rules of combination.*

Experiments on the flowering time of hybrids are not yet finished. However, it can already be reported that it is almost exactly intermediate between that of the seed plant and that of the pollen plant, and that development of the hybrids probably proceeds in the same manner with respect to this trait as it does for the remaining traits. The forms chosen for experiments of this nature must differ by at least 20 days in the mean date of blooming; it is also necessary that all the seeds be planted at an equal depth in order to achieve simultaneous germination, and, furthermore, that throughout the flowering period any relatively large temperature fluctuations with consequent acceleration or delay in blooming be taken into account. It is obvious that this experiment has various difficulties to overcome and demands great attention.

When we try to summarize briefly the results obtained, we find that those differing traits that permit easy and certain differentiation in the experimental plants *show completely concordant behavior in hybrid association.* Half of the progeny of plants hybrid in one pair of differing traits are hybrid again, while the other half become constant, with the characteristics of the seed and pollen plants in equal proportion. When, through fertilization, several differing traits become united in a hybrid, its progeny represent the terms of a combination series, in which the series for any one pair of differing traits are combined.

The complete agreement shown by all characteristics tested probably permits and justifies the assumption that the same behavior can be attributed also to the traits which show less distinctly in the plants, and could therefore not be included in the individual experiments. An experiment on flower stems of different lengths gave on the whole a rather satisfactory result, although distinction and classification of the forms could not be accomplished with the certainty that is indispensable for correct experiments.

THE REPRODUCTIVE CELLS OF HYBRIDS

The results of the previously cited investigations suggested further experiments whose outcome would throw light on the composition of seed and pollen cells in hybrids. An important clue is the fact that in *Pisum* constant forms appear among the progeny of hybrids and that they do so in all combinations of the associated traits. In our experience we find everywhere confirmation that constant progeny can be formed only when germinal cells and fertilizing pollen are alike, both endowed with the potential for creating identical individuals, as in normal fertilization of pure strains. Therefore we must consider it inevitable that in a hybrid plant also identical factors are acting together in the production of constant forms. Since the different constant forms are produced in a *single* plant, even in just a *single* flower, it seems logical to conclude that in the ovaries of hybrids as many kinds of germinal cells (germinal vesicles), and in the anthers as many kinds of pollen cells are formed as there are possibilities for *constant* combination forms and that these germinal and pollen cells correspond in their internal make-up to the individual forms.

Indeed, it can be shown theoretically that this assumption would be entirely adequate to explain the development of hybrids in separate generations if one could assume at the same time that the different kinds of germinal and pollen cells of a hybrid are produced on the average in equal numbers.

In order to test this hypothesis experimentally, the following experiments were chosen. Two forms which differed constantly in seed shape and albumen coloration were combined by fertilization.

If the differing traits are again designated by *A*, *B*, *a*, *b*, then:

AB seed plant,	*ab* pollen plant,
A shape round,	*a* shape angular,
B albumen yellow,	*b* albumen green.

The artificially fertilized seeds were sown, together with several seeds of the two parental plants, and the most vigorous specimens were chosen for reciprocal crosses. Fertilized were:

1. The hybrid with pollen from *AB*.
2. The hybrid " " " *ab*.
3. *AB* " " " the hybrid.
4. *ab* " " " the hybrid.

For each of these four experiments all flowers on three plants were fertilized. If the above assumption were correct, then germinal and pollen cells of the forms *AB*, *Ab*, *aB*, *ab* should develop in the hybrids, and combined would be:

1. Germinal cells *AB*, *Ab*, *aB*, *ab* with pollen cells *AB*.
2. " " *AB*, *Ab*, *aB*, *ab* " " " *ab*.
3. " " *AB* " " " *AB*, *Ab*, *aB*, *ab*.
4. " " *ab* " " " *AB*, *Ab*, *aB*, *ab*.

From each of these experiments, therefore, only the following forms could result:

1. *AB*, *ABb*, *AaB*, *AaBb*.
2. *AaBb*, *Aab*, *aBb*, *ab*.
3. *AB*, *ABb*, *AaB*, *AaBb*.
4. *AaBb*, *Aab*, *aBb*, *ab*.

Furthermore, if the individual forms of the hybrid's germinal and pollen cells are, on the average, formed in equal numbers, then in each experiment the four combinations listed must stand in equal ratio to each other. Complete agreement of the numerical values was not to be expected, however, since in every fertilization, even in a normal one, some germinal cells fail to develop or die later, and even some well-developed seeds do not succeed in germinating after being planted. Moreover, the assumption requires only that there be a tendency to approach equality in the number of different kinds of germinal and pollen cells produced, yet this number does not have to be attained by each hybrid with mathematical exactness.

The primary objective of the *first* and *second* experiments was to test the composition of hybrid germinal cells; that of the *third* and *fourth* experiments was to determine the composition of pollen cells. As shown by the above compilation, the first and third experiments, like the second and fourth, had to give identical combinations. The results should have been partly observable in the second year in the shape and coloration of the artificially fertilized seeds. In the first and third experiments the dominating traits of shape and color, A and B, occur in every combination, constant in some, in hybrid association with the recessive characters *a* and *b* in others, and must, for that reason, imprint their characteristics on all seeds. Therefore, if the premise were correct, all seeds should appear round and yellow. In the second and fourth experiments, on the other hand, one association is hybrid in shape and color, and therefore the seeds are round and yellow; another is hybrid in shape and constant in the recessive trait of color, therefore the seeds are round and green. The third association is constant in the recessive trait of shape and hybrid in color, therefore the seeds are angular and yellow. The fourth is constant in both recessive traits; therefore the seeds are angular and green. Hence, one could expect four different kinds of seeds from these two experiments, namely: round yellow, round green, angular yellow, angular green. The yield was in complete agreement with these expectations. There were obtained in the

First experiment, 98 exclusively round yellow seeds;
Third " 94 " " " " ;

Second experiment, 31 round yellow, 26 round green, 27 angular yellow, 26 angular green seeds;

Fourth experiment, 24 round yellow, 25 round green, 22 angular yellow, 27 angular green seeds.

A favorable result could hardly be doubted any longer, but the next generation would have to provide the final decision. From the seeds sown, in the first experiment 90 plants, and in the third 87 plants, bore fruit in the following year; these yielded in

Experiments			
1.	*3.*		
20	25	round yellow seeds	*AB.*
23	19	round yellow and green seeds	*ABb.*
25	22	round and angular yellow seeds . . .	*AaB.*
22	21	round and angular, yellow and green seeds .	*AaBb.*

In the second and fourth experiments the round yellow seeds produced plants with round and angular, yellow and green seeds . . . . *AaBb.*

From the round green seeds plants were obtained with round and angular green seeds . . . . *Aab.*

The angular yellow seeds produced plants with angular yellow and green seeds . . . . *aBb.*

From the angular green seeds plants were raised which again yielded only angular green seeds . . . . *ab.*

Though in these two experiments also some seeds did not germinate, no change could be effected in the figures found in the preceding year, since each kind of seed produced plants that were, with respect to their seeds, alike among themselves and different from the others. Thus in the

Second experiment	*Fourth experiment*					
31	24	plants	yielded	seeds	of form	*AaBb.*
26	25	"	"	"	"	*Aab.*
27	22	"	"	"	"	*aBb.*
26	27	"	"	"	"	*ab.*

In all experiments, therefore, all forms postulated by the preceding hypothesis appeared, and did so in nearly equal numbers.

In a further test the traits of *flower color* and *stem length* were included in the experiments and selection was made in such a way that in the third year of experimentation every trait had to appear in *half* of all plants if the above assumption were correct. A, *B*, *a*, *b* serve again as designation of the different traits.

A flowers purplish-red, *a* flowers white.
B stem long, *b* stem short.

Form A*b* was fertilized by *ab*, producing hybrid A*ab*. In addition, *aB* was also fertilized with *ab*, yielding hybrid *aBb*. For further fertilization in the second year the hybrid A*ab* was used as seed plant, the hybrid *aBb* as pollen plant.

Seed plant A*ab*, Pollen plant *aBb*.
Possible germinal cells A*b*, *ab*, Pollen cells *aB*, *ab*.

From fertilization involving the possible germinal and pollen cells, four combinations had to result, namely:

$$AaBb + aBb + Aab + ab.$$

From this it becomes apparent that, according to the above assumption, of all plants in the third year of experimentation

Half should have	violet-red flowers (A*a*)	. .	terms	1, 3
" " "	white flowers (*a*)	. . .	"	2, 4
" " "	a long stem (*Bb*)	. . .	"	1, 2
" " "	a short stem (*b*)	. . .	"	3, 4

Out of 45 fertilizations of the second year, 187 seeds were obtained, from which 166 plants reached the flowering stage in the third year. Among them the individual terms appeared in the following numbers:

Term	*Color of flower*	*Stem*	
1	violet-red	long	47 times
2	white	long	40 "
3	violet-red	short	38 "
4	white	short	41 "

Therefore,

violet-red flower color	(*Aa*)	occurred in	85 plants
white " "	(*a*)	" "	81 "
long stem	(*Bb*)	" "	87 "
short "	(*b*)	" "	79 "

In this experiment, too, the proposed hypothesis finds adequate confirmation.

Experiments on a small scale were also made on the traits of *pod shape, pod color,* and *flower position,* and the results obtained were in full agreement: all combinations possible through union of the different traits appeared when expected and in nearly equal numbers.

Thus experimentation also justifies the assumption *that pea hybrids form germinal and pollen cells that in their composition correspond in equal numbers to all the constant forms resulting from the combination of traits united through fertilization.*

The difference of forms among the progeny of hybrids, as well as the ratios in which they are observed, find an adequate explanation in the principle just deduced. The simplest case is given by the series for *one pair of differing traits.* It is known that this series is described by the expression: $A + 2Aa + a$, in which A and *a* signify the forms with constant differing traits, and *Aa* the form hybrid for both. The series contains four individuals in three different terms. In their production, pollen and germinal cells of form A and *a* participate, on the average, equally in fertilization; therefore each form manifests itself twice, since four individuals are produced. Participating in fertilization are thus:

$$\text{Pollen cells } A + A + a + a$$
$$\text{Germinal cells } A + A + a + a$$

It is entirely a matter of chance which of the two kinds of pollen combines with each single germinal cell. However, according to the laws of probability, in an average of many cases it will always happen that every pollen form A and *a* will unite equally often with every germinal-cell form A and *a*; therefore, in fertilization, one of the two pollen cells A will meet a germinal cell A, the other a germinal cell *a*, and equally, one pollen cell *a* will become associated with a germinal cell A, the other with *a*.

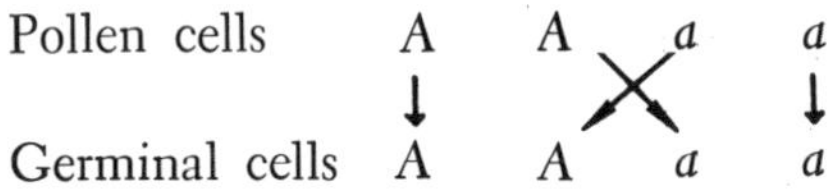

Pollen cells	A	A	*a*	*a*
	↓	↘	↙	↓
Germinal cells	A	A	*a*	*a*

The result of fertilization can be visualized by writing the designations for associated germinal and pollen cells in the form of fractions, pollen cells above the line, germinal cells below. In the case under discussion one obtains:

$$\frac{\mathrm{A}}{\mathrm{A}} + \frac{\mathrm{A}}{a} + \frac{a}{\mathrm{A}} + \frac{a}{a}$$

In the first and fourth terms germinal and pollen cells are alike; therefore the products of their association must be constant, namely A and *a*; in the second and third, however, a union of the two differing parental traits takes place again, therefore the forms arising from such fertilizations are absolutely identical with the hybrid from which they derive. *Thus, repeated hybridization takes place.* The striking phenomenon, that hybrids are able to produce, in addition to the two parental types, progeny that resemble themselves is thus explained: $\frac{\mathrm{A}}{a}$ and $\frac{a}{\mathrm{A}}$ both give the same association, A*a*, since, as mentioned earlier, it makes no difference to the consequence of fertilization which of the two traits belongs to the pollen and which to the germinal cell. Therefore

$$\frac{\mathrm{A}}{\mathrm{A}} + \frac{\mathrm{A}}{a} + \frac{a}{\mathrm{A}} + \frac{a}{a} = \mathrm{A} + 2\mathrm{A}a + a.$$

This represents the *average* course of self-fertilization of hybrids when two differing traits are associated in them. In individual flowers and individual plants, however, the ratio in which the members of the series are formed may be subject to not insignificant deviations. Aside from the fact that the numbers in which both kinds of germinal cells occur in the ovary can be considered equal only on the average, it remains purely a matter of chance which of the two kinds of pollen fertilizes each individual germinal cell. Therefore, isolated values must necessarily be subject to fluctuations, and even extreme cases are possible, as mentioned earlier in experiments on seed shape and albumen coloration. The true ratios can be given only by the mean calculated from the sum of as many separate values as possible; the larger their number the more likely it is that mere chance effects will be eliminated.

The series for hybrids in which *two kinds of differing traits* are associated contains 16 individuals representing 9 different forms, namely: $AB + Ab + aB + ab + 2ABb + 2aBb + 2AaB + 2Aab + 4AaBb$. Among the different traits of the parental plants A, a and B, b, 4 constant combinations are possible; therefore the hybrid produces the 4 corresponding forms of germinal and pollen cells, AB, Ab, aB, ab, and each of these will fertilize or be fertilized 4 times on the average, since the series contains 16 individuals. Participating in fertilization are thus

Pollen cells: $AB + AB + AB + AB + Ab + Ab + Ab + Ab + aB + aB + aB + aB + ab + ab + ab + ab.$

Germinal cells: $AB + AB + AB + AB + Ab + Ab + Ab + Ab + aB + aB + aB + aB + ab + ab + ab + ab.$

In fertilization every pollen cell unites, on the average, equally often with each form of germinal cell; thus each of the 4 pollen cells AB once with each of the germinal cell forms *AB*, *Ab*, *aB*, *ab*. In precisely the same manner the union of the remaining pollen cells of types *Ab*, *aB*, *ab* with all the other germinal cells takes place. Thus one obtains:

$$\frac{AB}{AB}+\frac{AB}{Ab}+\frac{AB}{aB}+\frac{AB}{ab}+\frac{Ab}{AB}+\frac{Ab}{Ab}+\frac{Ab}{aB}+\frac{Ab}{ab}+$$

$$\frac{aB}{AB}+\frac{aB}{Ab}+\frac{aB}{aB}+\frac{aB}{ab}+\frac{ab}{AB}+\frac{ab}{Ab}+\frac{ab}{aB}+\frac{ab}{ab},$$

or

$$AB + ABb + AaB + AaBb + ABb + Ab + AaBb + Aab + AaB + AaBb + aB + aBb + AaBb + Aab + aBb + ab$$
$$= AB + Ab + aB + ab + 2ABb + 2aBb + 2AaB + 2Aab + 4AaBb.$$

The series of hybrids in which *three kinds of differing traits* are combined can be explained in quite similar fashion. The hybrid produces 8 different forms of germinal and pollen cells: *ABC*, *ABc*, *AbC*, *Abc*, *aBC*, *aBc*, *abC*, *abc*, and again each pollen form unites once, on the average, with each germinal-cell form.

The law of combination of differing traits according to which hybrid development proceeds thus finds its basis and explanation in the proven statement that hybrids produce germinal and pollen cells that correspond in equal numbers to all the constant forms resulting from the combination of traits united through fertilization.

EXPERIMENTS ON HYBRIDS OF OTHER PLANT SPECIES

The object of further experiments will be to determine whether the law of development discovered for *Pisum* is also valid for hybrids of other plants. Several experiments were started quite recently for this purpose. I have completed two fairly small experiments with species of *Phaseolus,* which might be mentioned here.

An experiment with *Phaseolus vulgaris* and *Phaseolus nanus* L. gave fully concordant results. *Ph. nanus,* in addition to a dwarf-like stem, had green smoothly arched pods; *Ph. vulgaris,* on the other hand, had a stem 10–12′ long and yellow pods, constricted at maturity. The numerical relationships in which different forms occurred in individual generations were the same as in *Pisum.* The formation of constant associations also proceeded according to the law of simple combination of traits, exactly as in *Pisum.* Obtained were:

Constant combination	*Stem*	*Color of unripe pod*	*Shape of ripe pod*
1	long	green	arched
2	"	"	constricted
3	"	yellow	arched
4	"	"	constricted
5	short	green	arched
6	"	"	constricted
7	"	yellow	arched
8	"	"	constricted

The green pod color, arched pod shape, and tall stem were dominating traits, as in *Pisum.*

Another experiment with two very different *Phaseolus* species was only partly successful. Serving as *seed plant* was *Ph. nanus* L., a very constant species with white blossoms in short racemes and small white seeds in straight, arched, smooth pods; as *pollen plant Ph. multiflorus* W. with tall winding stem, crimson blossoms in very long racemes, rough sickle-like crooked pods and large seeds with black flecks and splashes on a peachblossom-red background.

The hybrid bore the greatest resemblance to the pollen plant, but the blossoms seemed less intensely colored. Its fertility was very limited; from 17 plants that developed a total of many hundreds of blossoms, only 49 seeds were harvested. These were of medium size and bore a design similar to that of *Ph. multiflorus*; the background color also did not differ basically. In the following year they produced 44 plants of which only 31 reached the flowering stage. The traits of *Ph. nanus*, which all became latent in the hybrid, reappeared in various combinations; their proportion to the dominating traits, however, fluctuated greatly because of the small number of experimental plants; but for some traits, like stem and pod shape, it was, as in *Pisum*, almost exactly 1:3.

Limited as the results of this experiment may be for the determination of ratios in which the various forms occurred, yet it provides a case of a *remarkable color change* in the blossoms and seeds of hybrids. It is known that in *Pisum* the traits of blossom and seed color appear unchanged in the first and in later generations, and that the offspring of hybrids carry exclusively one or the other of the two parental traits. The situation is different in the present experiment. True, the white flower and seed color of *Ph. nanus* appeared immediately in the first generation on one fairly fertile plant, but the remaining 30 plants developed flower colors that represented several gradations from crimson to pale violet. The coloration of the seed pod was no less varied than that of the flower. No plant could be considered fully fertile: some set no fruit at all; in others fruit was produced only by the last blossoms and did not have time to ripen. Well-formed seeds were harvested from only 15 plants. The greatest tendency toward infertility appeared in predominantly red-flowering forms; out of 16 such plants only 4 yielded ripe seeds. Three of these had a seed pattern similar to that of *Ph. multiflorus*, but a more or less pale background color, the fourth plant yielded only one seed, of plain brown coloration. Forms with preponderantly violet flower color had dark-brown, black-brown, and totally black seeds.

The experiment was continued for two more generations under equally unfavorable conditions, since even among the progeny of fairly fertile plants there were again some that were poorly fertile or completely sterile. No flower and seed colors other than those mentioned appeared. Forms receiving one or more of the recessive traits in the first generation remained constant in those traits without exception. Also, among the plants with violet blossoms and brown or black seeds, a few showed no further change in flower and seed color in the next generation, but the majority yielded, in addition to identical offspring, some with white flowers and similarly colored seed coats. Red-flowering plants remained so poorly fertile that nothing can be said with certainty about their further development.

Despite the many obstacles with which the observations had to contend, this experiment still establishes that development of hybrids follows the same law as in *Pisum* with respect to those traits concerned with the shape of the plant. Concerning the color traits, however, it seems difficult to find sufficient agreement. Besides the fact that a union of white and crimson coloration produces a whole range of colors from purple to pale violet and white, it is also striking that out of 31 flowering plants only one received the recessive trait of white coloration, while in *Pisum* this is true of every fourth plant on the average.

But these puzzling phenomena, too, could probably be explained by the law valid for *Pisum* if one might assume that in *Ph. multiflorus* the color of flowers and seeds is composed of two or more totally independent colors that behave individually exactly like any other constant trait in the plant. Were blossom color A composed of independent traits $A_1 + A_2 + \ldots$, which produce the overall impression of crimson coloration, then, through fertilization with the differing trait of white color a, hybrid associations $A_1a + A_2a + \ldots$ would have to be formed; and the situation with the corresponding coloration of the seed coat would be similar. According to the above assumption, each of these hybrid color combinations would be independent, and, therefore, would develop entirely independently from the rest. Then it is easily seen that from the combination of the individual series a complete color range should result. If, for instance, $A = A_1 + A_2$, then the series that correspond to hybrids A_1a and A_2a are

$$A_1 + 2A_1a + a,$$
$$A_2 + 2A_2a + a.$$

The terms of these series can enter into 9 different combinations, each of which represents the designation for another color:

1	A_1	A_2	2	A_1a	A_2	1	A_2	a,
2	A_1	A_2a	4	A_1a	A_2a	2	A_2a	a,
1	A_1	a	2	A_1a	a	1	a	a

The numbers preceding the individual combinations indicate how many plants of corresponding coloration belong to the series. Since their sum is 16, all colors are distributed over each 16 plants on the average, but, as the series itself shows, in unequal proportions.

If color development really occurred in this manner, then the above-mentioned case of white blossom and seed-coat color appearing only once among 31 plants of the first generation would have an explanation. This coloration occurs only once in the series and, therefore, could be expressed only in every 16 plants, on the average; for three color traits once only even among 64 plants.

One must not forget, however, that the explanation attempted here rests on a mere supposition, with nothing more to commend it than the very incomplete results of the experiment just discussed. It would be a worthwhile task, though, to follow color development in hybrids further by similar experiments, because it is probable that through this approach we can learn to understand the extraordinary diversity in the *coloration of our ornamental flowers.*

Up to now, hardly more is known with certainty than that the flower color in most ornamental plants is an extremely variable trait. The opinion has often been expressed that, through cultivation, species stability is greatly upset or entirely shattered, and there is a strong inclination to describe the development of cultivated forms as devoid of rules and subject to chance; usually the coloration of ornamental plants is pointed out as a model of instability. However, it is not clear why mere transplantation into garden soil should have such thorough and persistent revolution in the plant organism as its consequence. No one would seriously want to maintain that plant development in the wild and in garden beds was governed by different laws. Here as well as there changes in the type must appear when living conditions are changed and when a species has the ability to adapt itself to the new environment. Granted willingly that cultivation favors the formation of new varieties and that by the hand of man many an alteration has been preserved which would have perished in nature, but nothing justifies the assumption that the tendency to form varieties is so extraordinarily increased that species soon lose all stability and their progeny diverge into an infinite number of extremely variable forms. If the change in living conditions were the sole cause of variability one could expect that those cultivated plants that have been grown through centuries under almost identical conditions should have regained stability. This is known not to be the case, for it is precisely among them that not only the most different but also the most variable forms are found. Only Leguminosae, such as *Pisum, Phaseolus,* and *Lens,* whose organs of fertilization are protected by the keel, represent notable exceptions. During more than 1000 years of cultivation under the most diversified conditions, numerous varieties have arisen, yet these maintain stability under constant living conditions, just as do species growing wild.

It remains more than probable that a factor that so far has received little attention is involved in the variability of cultivated plants. Various experiences force us to accept the opinion that our cultivated plants, with few exceptions, are *members of different hybrid series* whose development along regular lines is altered and retarded by frequent intraspecific crosses. It should not be overlooked that cultivated plants are usually raised in fairly large numbers in close proximity to each other, a condition most favorable for reciprocal fertilization among the varieties present and between the species themselves. The likelihood that this opinion is correct is supported by the fact that among the large array of variable forms one finds always some single ones that remain constant for one or the other trait if all extraneous influence is carefully excluded. These forms develop exactly like certain members of the composite hybrid series. Even with respect to the most sensitive of all traits, that of color, it cannot escape careful observation that a tendency to variability exists in the individual forms to a very different degree. Among plants originating from a *single* spontaneous fertilization there are frequently some whose progeny diverge widely in the type and disposition of colors, while others produce forms that deviate little, and, if the number of plants is fairly large, some are encountered that transmit the color of their flowers unchanged to their progeny. Cultivated *Dianthus* species are an instructive example of this. A white-flowering specimen of *Dianthus caryophyllus,* itself derived from a white-flowered variety, was isolated in a greenhouse during the flowering period; its numerous seeds grew into plants with flowers of exactly the same shade of white. A similar result was obtained from a red, slightly violet glistening sport and from a white one with red stripes. On the other hand, many others, protected in the same manner, produced more or less differently colored and patterned progeny.

Anyone surveying the shades of color that appear in ornamental plants as a result of like fertilization cannot easily escape the conviction that here, too, development proceeds according to a certain law which possibly finds its expression through the *combination of several independent color traits.*

CONCLUDING REMARKS

A comparison of the observations made on *Pisum* with the experimental results obtained by Kölreuter and Gärtner, the two authorities in this field, cannot fail to be of interest. Both concur in the opinion that, in external appearance, hybrids either maintain a form intermediate between the parental strains or they approach the type of one or the other, sometimes being barely distinguishable from them. Various forms that diverge from the normal type usually arise from the seeds of hybrids that were fertilized by their own pollen. As a rule the majority of individuals produced from such a fertilization maintain the form of the hybrid, a few become more like the seed plant, and an occasional individual very nearly matches the pollen plant. This, however, is not valid for all hybrids without exception. Among the offspring of certain individuals some are more like one original stock plant, some more like the other, or they all tend more to one side than the other; but those from a few remain *exactly like the hybrid* and propagate unchanged. The hybrids of varieties behave like species hybrids, but possess a still greater inconstancy and a more pronounced tendency to revert to the original forms.

With respect to the *features* of hybrids and their regular *development,* consistency with the observations made on *Pisum* is unmistakable. This is not so in the exceptional cases mentioned. Gärtner himself admits that precise determination of whether a form bears a greater resemblance to one or the other of the two parental types often presents great difficulties, since much depends on the subjective viewpoint of the observer. And there is yet another circumstance that could contribute to making the results variable and uncertain in spite of the most careful observation and discrimination. For the most part plants which are considered to be good species and that differ in a rather large number of traits were used in the experiments. When one is dealing in a general way with degrees of similarity, then account must be taken not only of the traits that stand out sharply, but also of those that are often difficult to put into words, yet, as everyone familiar with plants knows, are sufficiently pronounced to give such forms the appearance of a stranger. If it is assumed that development of hybrids follows the law valid for *Pisum*, then the series obtained in each separate experiment must comprise very many forms, because the number of terms is known to increase with the number of differing traits as a power of three. Thus with a relatively small number of experimental plants the result could be only approximately correct and occasionally could deviate not inconsiderably. If, for instance, the two original stocks differed in 7 traits, and if 100 to 200 plants were raised from the seeds of their hybrids for an evaluation of the offsprings' degree of relationship, we can easily understand how uncertain such judgment must be, since the series for 7 differing traits contains 16,384 individuals appearing in 2187 different forms. Sometimes one relationship, sometimes another, would assert itself more strongly, depending on whether the observer found, by chance, a larger number of this or of that form.

Furthermore, when the differing traits include *dominating* ones that are passed on to the hybrid totally or almost totally unchanged, then the one of the two parental types having the larger number of dominating traits must always be the more prominent among the members of the series. In the experiment with three differing traits in *Pisum* described earlier, all of the dominating characters belonged to the seed plant. Although the members of the series tend equally toward both original parents in their internal makeup, the appearance of the seed plant was so preponderant in this experiment that 54 plants out of every 64 in the first generation looked exactly like it, or differed from it in only one trait. One sees how risky it can sometimes be to draw conclusions about the internal kinship of hybrids from their external similarity.

Gärtner mentions that in cases where development was regular the two parental types themselves were not represented among the offspring of the hybrids, only occasional individuals closely approximating them. Indeed, it cannot be otherwise in very extensive series. For 7 differing traits, for instance, each parental form occurs only once in more than 16,000 offspring of the hybrid. Therefore there is not much likelihood of finding them among a small number of experimental plants, yet, with a reasonable degree of probability, one may count on the appearance of a few forms that approximate those in the series.

We encounter an *essential difference* in those hybrids that remain constant in their progeny and propagate like pure strains. According to Gärtner these include the *highly fertile* hybrids *Aquilegia atropurpurea-canadensis, Lavatera pseudolbia-thuringiaca, Geum urbano-rivale*, and some *Dianthus* hybrids; according to Wichura it includes the hybrids of willow species. This feature is of particular importance to the evolutionary history of plants, because constant hybrids attain the status of *new species*. The correctness of these observations is vouched for by eminent observers and cannot be doubted. Gärtner had the opportunity of following the *Dianthus Armeria-deltoides* to its tenth generation, since that plant propagated itself regularly in the garden.

It was proven experimentally that in *Pisum* hybrids form *different kinds* of germinal and pollen cells and that this is the reason for the variability of their offspring. For other hybrids whose offspring behave similarly, we may assume the same cause; on the other hand, it seems permissible to assume that the germ cells of those that remain constant are identical, and also like the primordial cell of the hybrid. According to the opinion of famous physiologists, propagation in phanerogams is initiated by the union of one germinal and one pollen cell into one single cell,[5] which is

[5] It is presumably beyond doubt that in *Pisum* a complete union of elements from both fertilizing cells has to take place for the formation of a new embryo. How else could one explain that both parental types recur in equal numbers and with all their characteristics in the off-

able to develop into an independent organism through incorporation of matter and the formation of new cells. This development proceeds in accord with a constant law based on the material composition and arrangement of the elements that attained a viable union in the cell. When the reproductive cells are of the same kind and like the primordial cell of the mother, development of the new individual is governed by the same law that is valid for the mother plant. When a germinal cell is successfully combined with a *dissimilar* pollen cell we have to assume that some compromise takes place between those elements of both cells that cause their differences. The resulting mediating cell becomes the basis of the hybrid organism whose development must necessarily proceed in accord with a law different from that for each of the two parental types. If the compromise be considered complete, in the sense that the hybrid embryo is made up of cells of like kind in which the differences are *entirely and permanently mediated,* then a further consequence would be that the hybrid would remain as constant in its progeny as any other stable plant variety. The reproductive cells formed in its ovary and anthers are all the same and like the mediating cell from which they derive.

One could perhaps assume that in those hybrids whose offspring are *variable* a compromise takes place between the differing elements of the germinal and the pollen cell great enough to permit the formation of a cell that becomes the basis for the hybrid, but that this balance between the antagonistic elements is only temporary and does

---

spring of hybrids? If the influence of the germinal cell on the pollen cell were only external, if it merely played the role of a foster mother, then the outcome of each artificial fertilization would have to be that the resulting hybrid resembled the pollen plant exclusively or very closely. Experiments have in no way confirmed this up to now. A thorough proof for complete union of the content of both cells presumably lies in the universally confirmed experience that it is immaterial to the form of the hybrid which of the parental types was the seed or pollen plant.

not extend beyond the lifetime of the hybrid plant. Since no changes in its characteristics can be noticed throughout the entire vegetative period, we must further conclude that the differing elements succeed in escaping from the enforced association only at the stage at which the reproductive cells develop. In the formation of these cells all elements present participate in completely free and uniform fashion, and only those that differ separate from each other. In this manner the production of as many kinds of germinal and pollen cells would be possible as there are combinations of potentially formative elements.

This attempt to relate the important difference in the development of hybrids to a *permanent or temporary association* of differing cell elements can, of course, be of value only as a hypothesis which, for lack of well-substantiated data, still leaves some latitude. Some justification for the opinions expressed lies in the proof cited here that in *Pisum* the behavior of a pair of differing traits in hybrid union is independent of any other differences between the two parental plants and that, furthermore, the hybrid produces as many kinds of germinal and pollen cells as there are possible constant combination forms. The distinguishing traits of two plants can, after all, be caused only by differences in the composition and grouping of the elements existing in dynamic interaction in their primordial cells.

Yet even the validity of the laws proposed for *Pisum* needs confirmation, and a repetition of at least the more important experiments is therefore desirable: for instance, the one on the composition of hybrid fertilizing cells. An individual observer can easily overlook a distinguishing point that seems unimportant in the beginning but can grow to such proportions that it may not be neglected in the final analysis. Whether variable hybrids of other plant species show complete agreement in behavior also remains to be decided experimentally; one might assume, however, that no basic difference could exist in important matters since *unity* in the plan of development of organic life is beyond doubt.

Finally, the experiments performed by Kölreuter, Gärtner, and others on *transformation of one species into another by artificial fertilization* deserve special mention. Particular importance was attached to these experiments; Gärtner counts them as among "the most difficult in hybrid production."

When species *A* was to be transformed into *B*, the two were combined by fertilization and the resulting hybrids once more fertilized with pollen from *B*; from among their various descendants those closest to species *B* were then chosen and repeatedly fertilized by pollen from *B*, and so on, until finally a form that was like *B* and remained constant in its progeny was obtained. Thus species *A* was transformed into the other species, *B*. Gärtner himself has carried out 30 experiments of this kind with plants from genera *Aquilegia, Dianthus, Geum, Lavatera, Lychnis, Malva, Nicotiana,* and *Oenothera*. The length of time needed for transformation was not the same with all species. Although three successive fertilizations were sufficient for some, with others fertilizations had to be repeated five to six times; even with the same species fluctuations were observed in different experiments. Gärtner ascribes these differences to the circumstance that "the characteristic force toward change and transformation of the maternal type that a species exerts in reproduction is very different in different plants, and consequently the length of time required for one species to become transformed into another, and the number of generations it takes, must also be different; transformation is accomplished after more generations in some species, after fewer in others." The same observer notes further "that in the process of transformation much depends on which type and which individual was chosen for further transformation."

If one may assume that the development of forms proceeded in these experiments in a manner similar to that in *Pisum*, then the entire process of transformation would have a rather simple explanation. The hybrid produces as many kinds of germinal cells as there are constant combinations made possible by the traits associated within the hybrid, and one of these is always just like the fertilizing pollen cells. Thus there is the possibility that in such experiments a constant form identical to the pollen parent will result from the second fertilization. Whether one is actually obtained depends on the number of plants in each experiment as well as on the number of differing traits that were united by the fertilization. Let us assume, for example, that the plants chosen for the experiment differ in three traits and that species *ABC* is to be transformed into species *abc* by repeated fertilization with pollen from the latter. The hybrid resulting from the first fertilization forms 8 different kinds of germinal cells, namely:

ABC, ABc, AbC, aBC, Abc, aBc, abC, abc.

In the second year of the experiment these are again combined with pollen cells *abc* and one obtains the series:

$$AaBbCc + AaBbc + AabCc + aBbCc + Aabc + aBbc + abCc + abc.$$

Since the form *abc* occurs once in the series of 8 terms there is little likelihood that it would be missing among the experimental plants, even if only a fairly small number were raised, and transformation would thus be complete after two fertilizations. If, by chance, no transformation was obtained, fertilization would have to be repeated on one of the closest related combinations, *Aabc, aBbc, abCc*. It becomes obvious that *the smaller the number of experimental plants and the larger the number of differing traits* in the two parental species the longer an experiment of this kind will last, and that furthermore, a delay of one or even two generations could easily occur with these same species, which is what Gärtner has observed. The transformation of widely divergent species cannot be completed before the fifth or sixth experimental year because the number of different germinal cells formed in the hybrid increases with the number of differing traits as a power of two.

Gärtner found by repeated experiments that the *reciprocal* period of transformation varies for some species, so that quite frequently species *A* can be transformed into species *B* a generation earlier than species *B* into species A. From this he deduces that Kölreuter's opinion that "the two natures in hybrids are in perfect equilibrium" is not entirely tenable. It seems, however, that Kölreuter does not deserve this reproach but rather that Gärtner has overlooked an important point, to which he himself draws attention elsewhere, namely, that it "depends on which individual is chosen for further transformation." Experiments set up for this purpose with two *Pisum* species indicate that in the selection of individuals best suited for the purpose of further fertilization it could make a great difference which of the two species is to be transformed into the other. The two experimental plants differed in five traits; those of species *A* were all dominating, those of species *B* were all recessive. To effect mutual transformation *A* was fertilized with pollen from *B* and *B* with pollen from A, and the same procedure was repeated on both hybrids in the following year. In the first experiment, $\frac{B}{A}$, 87 plants *in the* 32 *possible forms* were available in the third experimental year from which to choose individuals for further fertilization; in the second experiment, $\frac{A}{B}$, the external appearance of all 73 plants obtained completely *coincided with that of the pollen plant*, although their internal constitution must have been just as varied as the forms from the other experiment. Intentional selection was therefore possible only in the first experiment; in the second one a few plants had to be chosen purely at random. Only a few flowers of the latter were fertilized with pollen from A, the rest were allowed to self-fertilize.

Among each five plants used for fertilization in the two experiments, the next year's culture showed the following agreement with the pollen plant:

First Experiment	*Second Experiment*	
2 plants	——	In all traits
3 "	——	" 4 "
——	2 plants	" 3 "
——	2 "	" 2 "
——	1 plant	" 1 trait

In the first experiment transformation was thus completed; in the second, which was not continued, two more fertilizations would probably have been necessary.

Though it is infrequently true that the dominating traits belong exclusively to one or the other parental plant, it will always make a difference *which* of the two possesses them in larger number. When the pollen plant has the majority of dominating traits, the choice of forms for further fertilization will afford a lesser degree of certainty than in the opposite case. A delay in the length of time needed for transformation will be the consequence if the experiment be considered complete only when a form is obtained that not only resembles the pollen plant in appearance but, like it, remains constant in its progeny.

The success of transformation experiments led Gärtner to disagree with those scientists who contest the stability of plant species and assume continuous evolution of plant forms. In the complete transformation of one species into another he finds unequivocal proof that a species has fixed limits beyond which it cannot change. Although this opinion cannot be adjudged unconditionally valid, considerable confirmation of the earlier expressed conjecture on the variability of cultivated plants is to be found in the experiments performed by Gärtner.

Among the experimental species were cultivated forms

such as *Aquilegia atropurpurea* and *canadensis, Dianthus Caryophyllus, chinensis* and *japonicus, Nicotiana rustica* and *paniculata,* and these, too, lost none of their stability after 4 to 5 repetitions of hybrid association.

On Hieracium-Hybrids Obtained By Artificial Fertilisation

Uber einige aus künstlicher Befruchtung gewonnenen Hieracium-Bastarde

GREGOR MENDEL

Communicated to the Meeting 9 June, 1869

Translation from Bateson, 1902[1]

Although I have already undertaken many experiments in fertilisation between species of *Hieracium*, I have succeeded in obtaining only the following 6 hybrids, and only from one to three specimens of them:

H. Auricula ♀ + *H. aurantiacum* ♂ [2]
H. Auricula ♀ + *H. Pilosella* ♂
H. Auricula ♀ + *H. pratense* ♂
H. echioides[3] ♀ + *H. aurantiacum* ♂
H. praealtum ♀ + *H. flagellare* Rchb. ♂
H. praealtum ♀ + *H. aurantiacum* ♂

[The original paper was published in Verhandlungen des naturforschenden Vereines in Brünn 8 (1869), Abhandlungen, p. 26ff; it appeared in 1870.]

[1] [Minor corrections have been made in the translation by the editors of this volume.]

[2] This connotation indicates that the hybrid was obtained by fertilization of *H. Auricula* with pollen from *H. aurantiacum*.

[3] The plant used in this experiment is not exactly the typical

The difficulty of obtaining a larger number of hybrids is due to the minuteness of the flowers and their peculiar structure. On account of this circumstance it was seldom possible to remove the anthers from the flowers chosen for fertilisation without either letting pollen get on to the stigma or injuring the pistil so that it withered away. As is well known, the anthers are united to form a tube, which closely embraces the pistil. As soon as the flower opens, the stigma, already covered with pollen, protrudes. In order to prevent self-fertilisation the anther-tube must be taken out before the flower opens, and for this purpose the bud must be slit up with a fine needle. If this operation is attempted at a time when the pollen is mature, which is the case two or three days before the flower opens, it is seldom possible to prevent self-fertilisation; for with every care it is not easily possible to prevent a few pollen grains getting scattered and communicated to the stigma. No better result has been obtained hitherto by removing the anthers at an earlier stage of development. Before the approach of maturity the tender pistil and stigma are exceedingly sensitive to injury, and even if they are not actually injured, they generally wither and dry up after a little time if deprived of their protecting investments. I hope to obviate this last misfortune by placing the plants after the operation for two or three days in the damp atmosphere of a greenhouse. An experiment lately made with *H. Auricula* treated in this way gave a good result.

To indicate the object with which these fertilisation experiments were undertaken, I venture to make some preliminary remarks respecting the genus *Hieracium*. This genus possesses such an extraordinary profusion of distinct forms that no other genus of plants can compare with it. Some of these forms are distinguished by special peculiarities and

---

H. echioides. It appears to belong to the series transitional to *H. praealtum*, but approaches more nearly to *H. echioides* and for this reason was reckoned as belonging to the latter.

may be taken as type-forms or species, while all the rest represent intermediate or transitional forms by which the type-forms are connected together. The difficulty in the separation and delimitation of these forms has demanded the close attention of the experts. Regarding no other genus has so much been written or have so many and such fierce controversies arisen, without as yet coming to a definite conclusion. It is obvious that no general understanding can be arrived at, so long as the value and significance of the intermediate and transitional forms are unknown.

Regarding the question whether and to what extent hybridisation plays a part in the production of this wealth of forms, we find various and conflicting views held by leading botanists. Although some of them maintain that this phenomenon has a far-reaching influence, others—for example, Fries—will have nothing to do with hybrids in *Hieracia*. Others take an intermediate position; and while granting that hybrids between the species in a wild state are not rare, still maintain that no great importance is to be attached to the fact, on the ground that they are only of short duration. The [suggested] causes of this are partly their restricted fertility or complete sterility; partly also the knowledge, obtained by experiment, that in hybrids self-fertilisation is always prevented if pollen of one of the parent-forms reaches the stigma. On these grounds it is regarded as inconceivable that *Hieracium* hybrids can constitute and maintain themselves as fully fertile and constant forms when growing near their progenitors.

The question of the origin of the numerous and constant intermediate forms has recently acquired no small interest since a famous *Hieracium* specialist has, in the spirit of the Darwinian teaching, defended the view that these forms are to be regarded as [arising] from the transmutation of lost or still-existing species.

From the nature of the subject it is clear that without an exact knowledge of the structure and fertility of the hybrids and the condition of their offspring through several generations no one can undertake to determine the possible influence exercised by hybridisation over the multiplicity of intermediate forms in *Hieracium.* The condition of the *Hieracium* hybrids in the range we are concerned with must necessarily be determined by experiments; for we do not possess a complete theory of hybridisation, and we may be led into erroneous conclusions if we take rules deduced from observation of certain other hybrids to be Laws of hybridisation, and try to apply them to *Hieracium* without further consideration. If by the experimental method we can obtain a sufficient insight into the phenomenon of hybridisation in *Hieracium,* then by the help of the information that has been collected respecting the structural relations of the wild forms, a satisfactory judgment in regard to this question may become possible.

Thus we may express the object which was sought after in these experiments. I venture now to relate the very slight results which I have as yet obtained with reference to this object.

1. Respecting the structure of the hybrids, we have to record the striking phenomenon that the forms hitherto obtained by similar fertilisation are not identical. The hybrids *H. praealtum* ♀ + *H. aurantiacum* ♂ and *H. Auricula* ♀ + *H. aurantiacum* ♂ are each represented by two, and *H. Auricula* ♀ + *H. pratense* ♂ by three individuals, while only one of each of the remainder has been obtained. If we compare the individual traits of the hybrids with the corresponding characters of the two parent types, we find that they sometimes present an intermediate structure, but are sometimes so near to one of the parent traits that the [corresponding character of the] other has receded considerably or almost evades observation. So, for instance, we see in one of the two forms of *H. Auricula* ♀ + *H. aurantiacum* ♂ pure yellow disc-florets; the petals of the marginal florets only are on the outside tinged with red to a scarcely noticeable degree: in the other, on the contrary, the colour of these florets comes very near to that of

H. aurantiacum, except that in the centre of the disc the orange red passes into a deep golden-yellow. This difference is noteworthy, for the flower-colour in *Hieracium* has the value of a constant character. Other similar cases are to be found in the leaves, the peduncles, &c.

If the hybrids are compared with the parent types as regards the sum total of their traits, then the two forms of *H. praealtum* ♀ + *H. aurantiacum* ♂ constitute approximately intermediate forms which, however, do not agree in certain traits. On the contrary in *H. Auricula* ♀ + *H. aurantiacum* ♂ and in *H. Auricula* ♀ + *H. pratense* ♂ we see the forms widely divergent, so that one of them is nearer to one parental type and the other nearer to the other, while in the case of the last-named hybrid there is still a third which is almost precisely intermediate between them.

The supposition is then forced on us that we have here only single terms in yet unknown series which may be formed by the direct action of the pollen of one species on the egg-cells of another.

2. With a single exception the hybrids in question form seeds capable of germination. *H. echioides* ♀ + *H. aurantiacum* ♂ may be described as fully fertile; *H. praealtum* ♀ + *H. flagellare* ♂ as fertile; *H. praealtum* ♀ + *H. aurantiacum* ♂ and *H. Auricula* ♀ + *H. pratense* ♂ as partially fertile; *H. Auricula* ♀ + *H. Pilosella* ♂ as slightly fertile; and *H. Auricula* ♀ + *H. aurantiacum* ♂ as infertile. Of the two forms of the last-named hybrid, the red-flowered one was completely sterile, but from the yellow-flowered one a single well-formed seed was obtained. Moreover, it must not pass unmentioned that among the seedlings of the partially fertile hybrid *H. praealtum* ♀ + *H. aurantiacum* ♂ there was one plant which possessed full fertility.

[3.] As yet the offspring produced by self-fertilisation of the hybrids have not varied, but agree in their traits both with each other and with the hybrid plant from which they were derived. From *H. praealtum* ♀ + *H. flagellare* ♂ two generations have flowered; from *H. echioides* ♀ + *H. aurantiacum*

♂, *H. praealtum* ♀ + *H. aurantiacum* ♂, *H. Auricula* ♀ + *H. Pilosella* ♂ one generation of each, comprising from 14 to 112 individuals, has flowered.

4. The fact must be declared that in the case of the fully fertile hybrid *H. echioides* ♀ + *H. aurantiacum* ♂ the pollen of the parent types was not able to prevent self-fertilisation, though it was applied in great quantity to the stigmas protruding through the anther-tubes when the flowers opened.

From two flower-heads treated in this way seedlings were produced resembling this hybrid plant. A very similar experiment, carried out this summer with the partially fertile *H. praealtum* ♀ + *H. aurantiacum* ♂ led to the conclusion that those flower-heads in which pollen of the parent type or of some other species had been applied to the stigmas developed a notably larger number of good seeds than those which had been left to self-fertilisation alone. The explanation of this result must be sought only in the circumstance that as a large part of the pollen-grains of the hybrid, examined microscopically, show a defective structure, a number of egg-cells capable of fertilisation do not become fertilised by their own pollen in the ordinary course of self-fertilisation.

It not rarely happens that in fully fertile species in the wild state the formation of the pollen fails, and in many anthers not a single good grain is developed. If in these cases seeds are nevertheless formed, such fertilisation must have been effected by foreign pollen. In this way hybrids may easily arise by reason of the fact that many forms of insects, notably the industrious hymenoptera, visit the flowers of *Hieracia* with great zeal and are responsible for the pollen, which easily sticks to their hairy bodies, reaching the stigmas of neighbouring plants.

From the few facts that I am able to contribute it will be evident that the work scarcely extends beyond its first inception. I must express some scruple in describing in this place an account of experiments just begun. But the conviction that the prosecution of the proposed experiments will demand a whole series of years, and the uncertainty whether it will be granted to me to bring them to a conclusion have determined me to make the present communication. By the kindness of Dr. Nägeli, the Munich Director, who was good enough to send me species which were wanting, especially from the Alps, I am in a position to include a larger number of forms in my experiments. I venture to hope to be able to contribute something more by way of extension and confirmation of the present account as early as next year.

If finally we compare the described results, still very uncertain, with those obtained by crosses made between forms of *Pisum*, which I had the honour of communicating in the year 1865,[4] we find a very real distinction. In *Pisum* the hybrids, obtained from the immediate crossing of two forms, all have the same type, but their posterity, on the contrary, are variable and follow a definite law in their variations. In *Hieracium* according to the present experiments the exactly opposite phenomenon seems to be exhibited. In describing the *Pisum* experiments it was remarked that there are also hybrids whose posterity do not vary, and that, for example, according to Wichura the hybrids of *Salix* reproduce themselves like pure species. In *Hieracium* we may take it we have a similar case. Whether from this circumstance we may venture to draw the conclusion that the polymorphism of the genera *Salix* and *Hieracium* is connected with the special condition of their hybrids is still an open question, which may well be raised but not as yet answered.

[4] Verhandlung des naturforschenden Vereines in Brünn 4, Abhandlungen, p. 3.

Gregor Mendel's Letters to Carl Nägeli: 1866-1873

Gregor Mendel's Briefe an Carl Nägeli: 1866–1873

Translated by Leonie Kellen Piternick and George Piternick

I

HIGHLY ESTEEMED SIR:

The acknowledged pre-eminence your honor[1] enjoys in the detection and classification of wild-growing plant hybrids makes it my agreeable duty to submit for your kind consideration the description of some experiments in artificial fertilization.

The experiments, which were made with different varieties of *Pisum,* resulted in the offspring of the hybrids forming

[These letters form the main body of a paper published by C. Correns under the above title in: Abhandlungen der Mathematisch-Physischen Klasse der Königlich Sächsischen Gesellschaft der Wissenschaften 29 (1905): 189–265. Reprinted *in* Carl Correns, Gesammelte Abhandlungen zur Vererbungswissenschaft aus periodischen Schriften 1899–1924. (Fritz v. Wettstein, ed.) Berlin, Julius Springer, 1924, pp. 1237–1281. The translation presented here, with some changes by the editors, was originally published in *Genetics* 35, no. 5, pt. 2 (1950): 1–29.]

[1] The German term "Ew. Wohlgeboren" has been translated "your honor" throughout, although the terms are not strictly equivalent. Trs.

curious series, the members of which in equal measure resembled the two original types. The presence of nonvariant intermediate forms, which occurred in each experiment, seems to deserve special attention. In the series for two and three differing traits, discussed in the monograph[2] (pp. 20–21), the notations for the constant forms have been placed first, since the terms are arranged according to their coefficients; but they receive a more correct position when the terms are placed according to their natural relationship to the two parental types, whereby that term which represents a hybrid in all traits and has, at the same time, the highest coefficient, will be placed exactly in the center.

The results which Gärtner obtained in his experiments are known to me; I have repeated his work and have reexamined it carefully to find, if possible, an agreement with those laws of development which I found to be true for my experimental plant. However, try as I would, I was unable to follow his experiments completely, not in a single case! It is very regrettable that this worthy man did not publish a detailed description of his individual experiments, and that he did not diagnose his hybrid types sufficiently, especially those resulting from like fertilizations. Statements like: "Some individuals showed closer resemblance to the maternal, others to the paternal type," or "the progeny had reverted to the type of the original maternal ancestor," etc., are too general, too vague, to furnish a basis for sound judgment. However, in most cases, it can at least be recognized that the possibility of an agreement with *Pisum* is not excluded. A decision can be reached only when new experiments in which the degree of kinship between the hybrid forms and their parental species are precisely determined, rather than simply estimated from general impressions, are performed.

In order to determine the agreement, if any, with *Pisum*, a study of those forms which occur in the first generation

[2] Mendel, Gregor [1866], Versuche über Pflanzenhybriden, Verh. des naturf. Vereines in Brünn 4 (1865): 3–47. The page numbers cited here refer to the translation of the paper found elsewhere in this volume.

should be sufficient. If, for two differing traits, the same ratios and series which exist in *Pisum* can be found, the whole matter would be decided. Isolation during the flowering period should not present many difficulties in most cases, since we are dealing only with few plants; those plants whose flowers are being fertilized and a few hybrids which have been selected for seed production. Those hybrids which are collected in the wild can be used as secondary evidence only, as long as their origin is not unequivocally known.

Hieracium, Cirsium, and *Geum* I have selected for further experiments. In the first two, manipulation in artificial pollination is very difficult and unreliable because of the small size and peculiar structure of the flowers. Last summer I tried to combine *H. Pilosella* with *pratense, praealtum,* and *Auricula;* and *H. murorum* with *umbellatum* and *pratense,* and I did obtain viable seeds; however, I fear that in spite of all precautions, self-fertilization did occur. The appearance of the young plants hardly suggests the desired result. *Hieracium* species can easily be grown in pots, and set abundant seed, even if they are confined in a room or a greenhouse during the flowering period.

In *Cirsium,* the dioeciously blooming *arvense* was fertilized by *oleraceum* and *canum.* The flowers were protected against visits of insects by coverings of bolting cloth; this protection appears to be sufficient for *Cirsium* species. Furthermore, the fertilization of *C. canum* and *C. lanceolatum* by *C. oleraceum* was attempted simply by transmission of pollen, without removing the anthers from the flowers of the former two. Whatever can be accomplished in the wild by insects should ultimately be possible by human hands, and among a great number of seedlings one should obtain a few hybrids. I plan to use the same procedure next summer with *Hieracium* as well.

The hybrid *Geum urbanum + rivale* deserves special attention. This plant, according to Gärtner, belongs to the few known hybrids which produce nonvariable progeny as long as they remain self-pollinated. To me it does not seem quite certain that the hybrid which Gärtner obtained was actually G. *intermedium* Ehrh. Gärtner calls his plant an intermediate type; this designation can not be applied without qualification to G. *intermedium*. In the transformation of G. *urbanum* into *rivale*, Gärtner states explicitly that, by fertilization of the hybrid with the pollen of *rivale*, only homogeneous offspring, which definitely resemble the paternal type, were obtained. However, we are not informed where this resemblance lies, and to what degree characters of G. *urbanum* were suppressed by each successive fertilization, until finally the pure *rivale* type emerged. It can hardly be doubted that this gradual transformation obeys a definite law, which, if it could be discovered, would also give clues to the behavior of other hybrids of this type. I hope to be able to get this artificial hybrid to flower next summer.

The surmise that some species of *Hieracium*, if hybridized, would behave in a fashion similar to *Geum*, is perhaps not without foundation. It is, for instance, very striking that the bifurcation of the stem, which must be considered transitional among the piloselloids, may appear as a perfectly constant trait, as I was able to observe last summer on seedlings of *H. stoloniflorum* W. K.

In the projected experiments with species of *Cirsium* and *Hieracium* I shall be entering a field in which your honor possesses the most extensive knowledge, knowledge that can be gained only through many years of zealous study, observation, and comparison of the manifold forms of these genera in their natural habitat. For the most part I lack this kind of experience because the press of teaching duties prevents me from getting into the field frequently, and during the vacations it is too late for many things. I am afraid that in the course of my experiments, especially with *Hieracium*, I shall encounter many difficulties, and therefore I am turning confidently to your honor with the request that you not deny me your esteemed interest when I need your advice.

With the greatest esteem and respect for your honor,

I subscribe myself,
GREGOR MENDEL
Monastery capitular and teacher in the high school

Brünn, 31 December 1866

II

HIGHLY ESTEEMED SIR:

My most cordial thanks for the printed matter you have so kindly sent me! The papers "Die Bastardbildung im Pflanzenreiche," "Über die abgeleiteten Pflanzenbastarde," "Die Theorie der Bastardbildung," "Die Zwischenformen zwischen den Pflanzenarten," "Die systematische Behandlung der Hieracien rücksichtlich der Mittelformen und des Umfanges der Species," especially capture my attention. This thorough revision of the theory of hybrids according to contemporary science was most welcome. Thank you again!

With respect to the essay which your honor had the kindness to accept, I think I should add the following information: the experiments which are discussed were conducted from 1856 to 1863. I knew that the results I obtained were not easily compatible with our contemporary scientific knowledge, and that under the circumstances publication of one such isolated experiment was doubly dangerous; dangerous for the experimenter and for the cause he represented. Thus I made every effort to verify, with other plants, the results obtained with *Pisum*. A number of hybridizations undertaken in 1863 and 1864 convinced me of the difficulty of finding plants suitable for an extended series of experiments, and that under unfavorable circumstances years might elapse without my obtaining the desired information. I attempted to inspire some control experiments, and for that reason discussed the *Pisum* experiments at the meeting of the local society of naturalists. I encountered, as was to be expected, divided opinion; however, as far as I know, no one undertook to repeat the experiments. When, last year [1866], I was asked to publish my lecture in the proceedings of the society, I agreed to do so, after having re-examined my records for the various years of experimentation, and not having been able to find a source of error. The paper which was submitted to you is the unchanged reprint of the draft of the lecture mentioned; thus the brevity of the exposition, as is essential for a public lecture.

I am not surprised to hear your honor speak of my experiments with mistrustful caution; I would not do otherwise in a similar case. Two points in your esteemed letter appear to be too important to be left unanswered. The first deals with the question whether one may conclude that constancy of type has been obtained if the hybrid *Aa* produces a plant A, and this plant in turn produces only A.

Permit me to state that, as an empirical worker, I must define constancy of type as the retention of a character during the period of observation. My statements that some of the progeny of hybrids breed true to type thus includes only those generations during which observations were made; it does not extend beyond them. For two generations all experiments were conducted with a fairly large number of plants. Starting with the third generation it became necessary to limit the numbers because of lack of space, so that, in each of the seven experiments, only a sample of those plants of the second generation (which either bred true or varied) could be observed further. The observations were extended over four to six generations (pp. 15–16). Of the varieties which bred true (pp. 18–21) some plants were observed for four generations. I must further mention the case of a variety which bred true for six generations, although the parental types differed in four traits. In 1859 I obtained a very fertile descendant with large, tasty, seeds from a first generation hybrid. Since, in the following year, its progeny retained the desirable characteristics and were uniform, the variety was cultivated in our vegetable garden, and many plants were raised every year up to 1865. The parental plants were *bcDg* and *BCdG*:

B albumen yellow	*b* albumen green
C seed coat grayish-brown	*c* seed coat white
D pod arched	*d* pod constricted
G stem long	*g* stem short

The hybrid just mentioned was *BcDG*.

The color of the albumen could be determined only in the plants saved for seed production, for the other pods were harvested in an immature condition. Never was green albumen observed in these plants, reddish-purple flower color (an indication of brown seed coat), constriction of the pod, nor short stem.

This is the extent of my experience. I cannot judge whether these findings would permit a decision as to constancy of type; however, I am inclined to regard the separation of parental traits in the progeny of hybrids in *Pisum* as complete, and thus permanent. The progeny of hybrids carries one or the other of the parental traits, or the hybrid form of the two; I have never observed gradual transitions between the parental traits or a progressive approach toward one of them. The course of development consists simply in this; that in each generation the two parental traits appear, separated and unchanged, and there is nothing to indicate that one of them has either inherited or taken over anything from the other. For an example, permit me to point to the packets, numbers 1035–1088, which I sent you. All the seeds originated in the first generation of a hybrid in which brown and white seed coats were combined. Out of the brown seed of this hybrid, some plants were obtained with seed coats of a pure white color, without any admixture of brown. I expect those to retain the same constancy of trait as found in the parental plant.

The second point, on which I wish to elaborate briefly, contains the following statement: "You should regard the numerical expressions as being only empirical, because they can not be proved rational."

My experiments with single traits all lead to the same result: that from the seeds of hybrids, plants are obtained half of which in turn carry the hybrid trait (Aa), the other half, however, receive the parental traits A and a in equal amounts. Thus, on the average, among four plants two have the hybrid trait Aa, one the parental trait A, and the other the parental trait a. Therefore $2Aa + A + a$ or $A + 2Aa + a$ is the empirical simple series for two differing traits. Likewise it was shown in an empirical manner that, if two or three differing traits are combined in the hybrid, the series is a combination of two or three simple series. Up to this point I don't believe I can be accused of having left the realm of experimentation. If then I extend this combination of simple series to any number of differences between the two parental plants, I have indeed entered the rational domain. This seems permissible, however, because I have proved by previous experiments that the development of a pair of differing traits proceeds independently of any other differences. Finally, regarding my statements on the differences among the ovules and pollen cells of the hybrids; they also are based on experiments. These and similar experiments on the germ cells appear to be important, for I believe that their results furnish the explanation for the development of hybrids as observed in *Pisum*. These experiments should be repeated and verified.

I regret very much not being able to send your honor the desired varieties. As I mentioned above, the experiments were conducted up to and including 1863; at that time they were terminated in order to obtain space and time for the growing of other experimental plants. Therefore seeds from those experiments are no longer available. Only one experiment on differences in the time of flowering was continued; and seeds are available from the 1864 harvest of this experiment. These are the last I collected, since I had to abandon the experiment in the following year because of devastation by the pea beetle, *Bruchus pisi*. In the early years of experimentation this insect was only rarely found on the plants, in 1864 it caused considerable damage, and appeared in such numbers in the following summer that hardly a 4th or 5th of the seeds was spared. In the last few years it has been necessary to discontinue cultivation of peas in the vicinity of Brünn. The seeds remaining can still be useful, among them are some varieties which I expect to remain constant; they are derived from hybrids in which two, three, and four differing traits are combined. All the seeds were obtained from members of the first generation, i.e., of such plants as were grown directly from the seeds of the original hybrids.

I should have scruples against complying with your honor's request to send these seeds for experimentation, were it not in such complete agreement with my own wishes. I fear that there has been partial loss of viability. Furthermore the seeds were obtained at a time when *Bruchus pisi* was already rampant, and I can not acquit this beetle of possibly transferring pollen; also, I must mention again that the plants were destined for a study of differences in flowering time. The other differences were also taken into account at the harvest, but with less care than in the major experiment. The legend which I have added to the packet numbers on a separate sheet is a copy of the notes I made for each individual plant, with pencil, on its envelope at the time of harvest. The dominating traits are designated as A, *B*, C, *D*, *E*, *F*, G and as concerns their dual meaning please refer to p. 13. The recessive traits are designated *a*, *b*, *c*, *d*, *e*, *f*, *g*; these should remain constant in the next generation. Therefore, from those seeds which stem from plants with recessive traits only, identical plants are expected (as regards the traits studied).

Please compare the numbers of the seed packets with those in my record, to detect any possible error in the designations—each packet contains the seeds of a single plant only.

Some of the varieties represented are suitable for experiments on the germ cells; their results can be obtained during the current summer. The round yellow seeds of packets 715, 730, 736, 741, 742, 745, 756, 757, and on the other hand, the green angular seeds of packets 712, 719, 734, 737, 749, and 750 can be recommended for this purpose. By repeated experiments it was proved that, if plants with green seeds are fertilized by those with yellow seeds, the albumen of the resulting seeds has lost the green color and has taken up the yellow color. The same is true for the shape of the seed. Plants with angular seeds, if fertilized by those with round or rounded seeds, produce round or rounded seeds. Thus, due to the changes induced in the color and shape of the seeds by fertilization with foreign pollen, it is possible to recognize the constitution of the fertilizing pollen.

Let *B* designate yellow color; *b*, green color of the albumen.

Let A designate round shape; *a*, angular shape of the seeds.

If flowers of such plants as produce green and angular seeds by self-fertilization are fertilized with foreign pollen, and if the seeds remain green and angular, then the pollen of the donor plant was, as regards the two traits . . . . *ab*

If the shape of the seeds is changed, the pollen was taken from . *Ab*

" " color " " " " " " " " " " . *aB*

" both shape and color " " " " " " " . *AB*

The packets enumerated above contain round and yellow, round and green, angular and yellow, and angular and green seeds from the hybrids *ab* + *AB*. The round and yellow seeds would be best suited for the experiment. Among them (see experiment p. 18) the varieties *AB*, *ABb*, *Aab*, and *AaBb* may occur; thus four cases are possible when plants, grown from green and angular seeds, are fertilized by the pollen of those grown from the above mentioned round and yellow seeds, i.e.

I. *ab* + *AB*
II. *ab* + *ABb*
III. *ab* + *AaB*
IV. *ab* + *AaBb*

If the hypothesis that hybrids form as many types of pollen cells as there are possible constant combination types is correct, plants of the makeup

AB	produce	pollen	of	the type	*AB*
ABb	"	"	"	"	*AB* and *Ab*
AaB	"	"	"	"	*AB* and *aB*
AaBb	"	"	"	"	*AB*, *Ab*, *aB*, and *ab*.

Fertilization of ovules occurs:

I.	Ovules	*ab*	with	pollen	*AB*
II.	"	*ab*	"	"	*AB* and *Ab*
III.	"	*ab*	"	"	*AB* and *aB*
IV.	"	*ab*	"	"	*AB*, *Ab*, *aB*, and *ab*.

The following varieties may be obtained from this fertilization:

I. *AaBb*
II. *AaBb* and *Aab*
III. *AaBb* and *aBb*
IV. *AaBb*, *Aab*, *aBb*, and *ab*.

If the different types of pollen are produced in equal numbers, there should be in

I. All seeds round and yellow
II. one half round and yellow
one half round and green

III. one half round and yellow
one half angular and yellow

IV. one quarter round and yellow
" " round and green
" " angular and yellow
" " angular and green.

Furthermore, since the numerical relations between *AB*, *ABb*, *AaB*, *AaBb* are 1:2:2:4, among any nine plants grown from round yellow seed there should be found on the average *AaBb* four times, *ABb* and *AaB* twice each, and *AB* once; thus case IV should occur four times as frequently as I and twice as frequently as II or III.

If on the other hand, plants grown from the round yellow seeds mentioned are fertilized by pollen from plants grown from green angular seeds, the results should be exactly the same, provided that the ovules are of the same types, and formed in the same proportions, as was reported for the pollen.

I have not performed this experiment myself, but I believe, on the basis of similar experiments, that one can depend on the result indicated.

In the same fashion individual experiments may be performed for each of the two seed traits separately; all those round seeds which occurred together with angular ones, and all the yellow ones which occurred with green seeds on the same plant are suitable. If, for instance, a plant with green seeds was fertilized by one with yellow seeds, the seeds obtained should be either (1) all yellow, or (2) half yellow and half green, since the plants originating from yellow seeds are of the varieties *B* and *Bb*. Since, furthermore, *B* and *Bb* occur in the ratio of 1:2, the second fertilization will occur twice as frequently as the first.

Regarding the other traits, the experiments may be conducted in the same way; results, however, will not be obtained until next year.

I have all the piloselloid *Hieracia* which your honor recommends for the experiments, also *H. murorum* and *H. vulgatum* of the *Archieracia; H. glaucum, H. alpinum, H. amplexicaule, H. prenanthoides,* and *H. tridentatum* do not occur in this vicinity. Last summer I found a withered *Hieracium,* which has the seed color of *prenanthoidea* (Fries: Achaenia typice testaces [pallida]), but did not resemble any of the herbarium specimens of this type very closely; finally our botanist declared it to be a hybrid. The rootstock has been transplanted to the garden for further observations, and the seeds have been planted. On the whole, this area is poor in *Hieracia,* and probably has not been sufficiently searched. Next summer I hope to have the time to roam the sandy lignite country which extends eastward from Brünn for several miles to the Hungarian frontier. Several other rare plants are known from this region. The Moravian plateau also is probably terra incognita as far as the *Hieracia* are concerned. If I should find anything noteworthy during the summer, I shall hurry to send it to your honor. At the moment permit me to include with the seed packets the plant just mentioned, albeit in a rather defective condition, together with another *Hieracium.* Last year I found at least 50 specimens of it on an old garden wall. This plant is not found in the local herbaria; its appearance suggests both *H. praealtum* and *H. echioides,* without being one or the other. *H. praealtum* does occur in the environs of the city, *H. echioides* does not.

Several specimens of the hybrid *Geum urbanum* + *G. rivale* (from last year's hybridization) wintered in the greenhouse. Three are now flowering, the others will follow. Their pollen is fairly well developed, and the plants should be fertile, just as Gärtner states. It seems strange that all the plants now flowering are of the exceptional type mentioned by Gärtner. He says: "*Geum urbano-rivale,* mostly with large flowers, like *rivale,* and only a few specimens with small yellow flowers like *urbanum.*" In my plants the flowers are yellow or yellow-orange, and about half the size of those of *G. rivale;* the other characters correspond, as far as can be judged at present, to those of G. *intermedium* Ehrh. Could it be that the exceptional type has an earlier flowering season? To judge from the buds, however, the other plants do not have large flowers either. Or could it be that the exception has become the rule? I believe I have good reasons for considering my parental species pure. I obtained G. *urbanum* in the environs of the city, where neither G. *rivale*, nor any other species of the genus occurs; and I got G. *rivale* in a damp mountain meadow, where G. *urbanum* certainly does not occur. This plant has all the characteristics of G. *rivale*; it is being maintained in the garden, and seedlings have been produced from self-fertilization.

The *Cirsium arvense* + C. *oleraceum* hybrids, sown in the fall, have died during the winter; one plant of the C. *arvense* + C. *canum* hybrid survived. I hope the spring seedlings will do better. Two other *Cirsium* hybrids have wintered well in the greenhouse. Last summer I observed, on a flowering plant of C. *praemorsum* M. (*olerac.* + *rivulare*), that in those heads which develop first and last on the stems, no pollen is formed, and thus they are completely sterile; on the others (about one half of the total heads) some pollen and fertile seed is formed. Fertilization experiments were conducted with two of the late-developing heads; pollen of C. *palustre* was transferred to one, pollen of C. *canum* to the other. Viable seeds were obtained from both, the resulting plants survived the winter in the greenhouse, and are now developed to a stage at which the success of the hybridization is evident. Some seedlings of C. *praemorsum*, others of a hybrid, probably in the group C. *canum* + *palustre*, and those of a third one, probably C. *rivulare* + *palustre*, have survived the winter in the open quite well. The same may be said of the autumn seedlings of the hybrids *Aquilegia canadensis* + *vulgaris*, A. *canadensis* + A. *atropurpurea*, and A. *canadensis* + A. *Wittmaniana*. However, fall seedlings of some *Hieracia* which were grown to test constancy of type have suffered considerable damage. In this genus it is preferable to sow in the early spring, but then it is doubtful that the plants will flower in the same year. Nevertheless, Fries has made this statement concerning the division Accipitrina: "Accipitrina, praecocius sata, vulgo primo anno florent."

I have obtained luxuriant plants of *Linaria vulgaris* + *L. purpurea;* I hope they will flower in the first year. The same may be said of *Calceolaria salicifolia* and *C. rugosa.* Hybrids of *Zea Mays major* (with dark red seeds) + *Z. Mays minor* (with yellow seeds) and of *Zea Mays major* (with dark red seeds) + *Zea Cuzko* (with white seeds) will develop during the summer. Whether *Zea Cuzko* is a true species or not I do not dare to state. I obtained it with this designation from a seed dealer. At any rate it is a very aberrant form. To study color development in flowers of hybrids, cross-fertilizations were made last year between varieties of *Ipomoea purpurea, Cheiranthus annuus,* and *Antirrhinum majus.* An experiment with hybrids of *Tropaeolum majus* + *T. minus* (first generation) must also be mentioned.

For the current year exploratory experiments with *Veronica, Viola, Potentilla,* and *Carex* are planned. Unfortunately, I have only a small number of species.

Because of lack of space the experiments can be started with a small number of plants only; after the fertility of the hybrids has been tested, and when it is possible to protect them sufficiently during the flowering period, each in turn will receive an extensive study. Thus far the three *Aquilegia* hybrids mentioned above and *Tropaeolum majus* + *T. minus* are suitable, although the latter has only partial fertility. It is hoped that *Geum urbanum* + *G. rivale* can be included in the group of suitable plants.

As must be expected, the experiments proceed slowly. In the beginning, some patience is required, but later, when several experiments are progressing concurrently, matters are improved. Every day, from spring to fall, one's interest is refreshed daily, and the care which must be given to one's wards is thus amply repaid. In addition, if I should, by my experiments, succeed in hastening the solution of these problems, I should be doubly happy.

Accept, highly esteemed Sir, the expression of most sincere respect from

Your devoted,
G. MENDEL
Altbrünn, Monastery of St. Thomas

Brünn, 18 April, 1867

III

HIGHLY ESTEEMED SIR:

My project of studying the *Hieracia* of this locality in their natural environment has, unfortunately, been carried out to only a very limited extent. Lack of time is chiefly to blame for this, and I am no longer very fit for botanical field trips, for heaven has blessed me with an excess of avoirdupois, which becomes very noticeable during long travels afoot, and, as a consequence of the law of general gravitation, especially when climbing mountains. If it thus not possible to send a collection of wild-growing *Hieracia,* as I should have liked to do, permit me to submit for your gracious attention some material from my garden instead.

First in importance as a *Hieracium* hybrid: *H. praealtum* + *H. stoloniflorum* (Autorum). I enclose the two parental plants for your critical examination, since I do not trust myself to make an accurate determination in this genus. It should be mentioned that this *H. Praealtum* (perhaps var. *obscurum* Rchb.) occurs frequently in somewhat damp localities such as meadows etc., in the vicinity of Brünn, and that it frequently grows more luxuriantly than does the cultivated specimen. Twenty-eight seedlings of the current year showed no variation. Runners are always absent. *H. stoloniflorum* also occurs locally, but only sporadically. I have raised this form from seeds obtained in Breslau from a herbarium specimen, to which the following note was added:

"Occurs commonly, and is not a hybrid." I can only state that the local plant is identical with that from Breslau, that the seedlings of this and last year (second generation) have shown no deviations, and that they are all completely fertile.

In addition to other experiments which I performed last year in order to achieve artificial fertilization in *Hieracia,* I also tried to inhibit pollen development in the above mentioned *H. praealtum,* or, at least, to prevent it from reaching the stigma. To this end, more than half the involucral bracts of a young, incompletely developed head were cut off, and the small flower buds, except for 10–12, removed; the latter were slit open in several places with a fine needle so as to expose the style completely. Fertilization with pollen of *H. stoloniflorum* was performed immediately and was repeated later on. In spite of this drastic treatment, four well-developed seeds were obtained, which, when sown in the spring, produced as many plants. Three were completely like *H. praealtum,* whereas the fourth showed considerable divergence, and doubtless represents the hybrid form *H. praealtum* + *H. stoloniflorum.* Thus at least once in four cases, self-fertilization had been prevented by the procedure described; the latter seems to be useful, although it is most complicated, and strains and tires the eye. Since, to judge from leaf formation in the young plant, fertilization of *Cirsium canum* with *C. oleraceum* was achieved in the same way last year, I have used the same procedure in all fertilizations among *Hieracia* this summer.

The *Hieracium* hybrid mentioned is a healthy, luxuriant plant. Early in July it developed simultaneously several vertical stems; runners were absent at the time. When the first heads were about to open, the plant with all its roots and surrounding earth was transplanted to a pot, and isolated while blooming. Only after withering of all heads did a short, thick, sterile runner appear, and soon root. The plant was later returned to the ground, and here it started to bloom a second time, at about the end of September; the stems, however, remained much shorter and weaker. Soon afterwards, a creeping, sterile stolon developed, which bore five vertical heads.

The hybrids *H. Pilosella* + *H. Auricula* and *H. Pilosella* + *H. praealtum*, produced by Fr. Schultz using artificial fertilization, were described as sterile. Since the hybrid *H. praealtum* + *H. stoloniflorum* furnished a number of good seeds, it deserves some attention. In a total of 14 heads 1044 flowers were counted; of these 624 furnished seeds good to all appearances; the majority, however, were not viable, since only 156 plants developed from them (about 15%). These have rooted well, and should flower next year. Whether they will retain the characteristics of the hybrid, or whether they will show variations, will be determined by next year's observations.

I should like to add some remarks on those traits of the hybrids which are rather difficult to determine in the dried parts.

The leaves are covered in the same fashion as those of *H. stoloniflorum*, but the bristles, especially those of the lower surface, are much less numerous; the stellate hairs are less dense. The stem is covered with stellate hairs, a few grayish-white bristles, and glandular hairs (in *H. praealtum* the base of the bristles is brown, the glandular hairs absent). Bracts of the involucre and flower stalks are densely covered with stellate and glandular hairs (bristles are lacking, as in *H. praealtum*). The sheath of the withered heads is only slightly inflated toward the stem; the marginal flowers are unicolored. On the average, the number of flowers per head is (according to 14 different counts) 39 in *praealtum*, 145 in *stoloniflorum*, and 75 in the hybrid. The latter number thus does not represent the mean, which should be 92; it is, however, almost the exact geometric mean; since 75^2 is approximately 39×145.

The plant will be observed again next year. Of this year's fertilizations among *H. Pilosella, stoloniflorum, Auricula,* and *praealtum* on the one hand, and among *H. murorum, vulgatum, rigidum, boreale,* and *umbellatum* on the other hand, I also hope to obtain more hybrids for further study.

Unfortunately I lack other species, necessary for my experiments. The seeds of the fertilized piloselloids were sown shortly after they matured, and the plants were transplanted to the garden, where they have developed rather well. To judge from the appearance of the basal rosette of leaves, the combination of *H. praealtum* (var. *Bauhini*) with *H. aurantiacum* may be considered successful. Planting of the *Archieracia* will take place next spring; they bloom 5–6 months after being sown.

Also included among the experimental plants was that species with pale brown seed of which I sent you a withered specimen last spring. The plant grows in the Punkwa Valley in the foothills of the Moravian mountains, on a lime-containing substrate, and is not uncommonly found on cut-over lands, in association with *H. murorum, H. vulgatum,* and *H. boreale*. In August I have found it starting to wither. In dry, infertile soil, the plant appears sparsely developed, and the lateral branches are poorly or not at all developed; in locations rich in humus, on the other hand, one can hardly recognize it as the same plant. I am enclosing the most luxuriant specimen I have found. Plants raised from seed in the garden show no variation, but grow even larger and more vigorously, although they will grow in unfertilized, sandy, soil, just as do other *Hieracia*.

Also included is a *Hieracium* (probably *praealtum*) whose seeds had a lighter color than the varieties of *H. praealtum* which are known to me. I collected the plant in the vicinity of Tscheitscher Lake, where it grows in abundance. On a sunny slope among low bushes I also found bifurcate types, which are probably of hybrid origin, in association with *H. praealtum* and *H. Pilosella.* I am enclosing some of them. The *H. praealtum* appears to be very close to *H. cymigerum* Rchb. The luxuriant specimen was collected in a somewhat damp location; the other on dry, stony, soil.

From last year's hybrid *Cirsium praemorsum + canum,* only two plants were obtained, and they bloomed this summer. I am sending leaves and flowers of both plants, which are designated as I and II respectively. The pinnatifid radical leaves of plant II were completely destroyed by snails; the inclosed leaves of this plant are cauline leaves at various levels. I must also mention that both plants were very luxuriantly developed, especially plant I. (Height, 6 feet, circumference of the stem at ground level, 6½ inches.) The lower cauline leaves were only slightly recurved, the upper ones not at all. Numerous branches develop from close to the base of the stem. The corollas, which are originally white, soon assume a yellowish-white color, and when withering, a straw-yellow color; the styles gradually turn red, to appear at last strongly crimson. In plant I, the lower half of the stem, the ribs of the radical leaves, and also parts of those of the cauline leaves, were of a dark red color; in plant II, this coloration was absent. Plant I bloomed early in July, plant II a whole month later. Plant I is of average fertility, plant II is nearly sterile. The roots can not be examined until next spring without damaging the plants. The striking differences in leaf formation and position of flowers may be recognized in the dried specimens. About 50 progeny of each parental plant are available from this year's spring sowing. Their development next summer will show whether or not variants do appear following self-fertilization, and what kinds, and what relation they bear to the differences between the two hybrids.

Fertilization of *C. praemorsum* with *C. canum* was repeated in order to obtain several hybrids for further comparisons.

Geum urbanum + *G. rivale* (from last year's hybridization) looks like *G. intermedium* Ehrh. The varieties with reddish-yellow flowers and those with half-size flowers do not occur among my hybrids. Not all of them had equal fertility, but none of the plants was completely sterile. Fertilized for further experiments were:

G. *urbanum* with the hybrid
G. *rivale* with the hybrid

the hybrid with G. *urbanum*
the hybrid with G. *rivale*

Plants obtained from these fertilizations, and those produced by self-fertilization from the hybrid were planted in the garden at the beginning of August.

The hybrid *Linaria vulgaris* + *L. striata* bloomed during the first year. I received the latter plant under the designation "*Linaria purpurea* (*Antirrhinum striatum*)"; it seems to be none other than *Linaria striata* DC. The hybrid is intermediate between the parental species as regards position of leaves and flowers, and size and shape of flowers; the fruits, however, are like those of *L. striata*; the inflated, wrinkled capsule form of *L. vulgaris* is not apparent in them. Flower color and shape of seed are peculiar. The bluish-violet striation of the upper lip especially is that of *L. striata*, the orange colored palate spot, that of *L. vulgaris*; the ground color of the flowers was pale yellow in 33 out of 55, pale violet in 21, and one plant produced both colors, but on separate flower stems. I am sending both flowers of the latter, dried in sand, as well as those of *L. striata* and G. *urbanum* + G. *rivale*. When flowers are dried between papers, much becomes indistinguishable.

The seeds of *L. vulgaris* are known to be flat and lenticular, with a rough surface, and equipped with a circular, broad, wing margin; those of *L. striata* are oval, sharply triangular, the surfaces wrinkled and pitted, and without wings. The seeds of the hybrid show considerable variation. While individual ones resemble those of *L. striata* very closely, being distinctly triangular, and lacking wings, the intermediate condition is suggested in the majority by an expansion of one of the three surfaces, while the opposite edge becomes rounded off, is merely suggested, or completely disappears in some seeds. In the latter case, the seeds get a plano-convex or concave-convex shape, and are equipped with a very narrow wing margin. Wrinkles and pits are always present, but less densely than in *L. striata*. Reciprocal fertilizations were undertaken in the same manner as in the *Geum* hybrid.

Linaria vulgaris may fairly easily be fertilized by the pollen of other *Linaria* species; among five attempts made this summer, four were successful. Among them is the combination with the beautiful *L. genistifolia*; this hybrid is reported to grow in the wild around Brünn. *L. vulgaris* could not be fertilized with *L. triphylla*.

To conclude, let me report an observation which I made last summer on a *Verbascum* hybrid. I undertook hybridizations among several species of *Verbascum* in 1864. The hybrids raised in the garden were completely sterile; not even a single seed was obtained. By accident one plant of the cross *Verbascum phoeniceum* + V. *Blattaria* was left in the seed bowl, and remained therein in a corner of the garden without any care throughout the summer. This very stunted plant was discovered in the autumn and planted in the open ground, along with its luxuriantly developing siblings. Although it became fairly vigorous during the following year, it did not bloom, and wintered for the second time, whereas its siblings died as two-year-old plants, after having bloomed. This summer the plant made up for everything it had missed, for from June to September it flowered continuously, and produced more than 100 well-formed seeds. It might winter for a third time, since a complete leaf rosette developed after flowering.

I am impatiently awaiting the coming summer, because several fertile hybrids will exhibit their progeny in flower for the first time. Care has been taken that they may appear in large numbers, and I only hope that they will reward my anticipation with much information about their life history.

With greatest respect I sign myself your honor's

Sincere admirer,
GREGOR MENDEL

Brünn, 6 Nov. 1867

IV

HIGHLY ESTEEMED SIR:

After having in the past two years collected some experience in the artificial fertilization of *Hieracia,* I intend to perform some systematic experiments with this genus, experiments which will be limited to crosses between the main types. I have, with few exceptions, the typical species of the piloselloids, but I lack almost all of the *Archieracia.* I should like to purchase the missing material, but where? That is the question I can not answer. In this predicament I dare to appeal to your honor's kindness, hoping to obtain the desired information.

The species which I should like to have are: *H. cymosum (genuinum), H. alpinum, H. amplexicaule, H. glanduliferum, H. piliferum, H. villosum, H. glaucum, H. porrifolium, H. humile, H. tridentatum, H. praenanthoides, H. albidum.*

H. glaciale, H. alpicola, and *H. staticifolium* would also be very welcome. I am very anxious to receive seeds, plants, or both, of the species enumerated. One might still expect plants to flower this summer from seeds sown now.

Repeating my request, I gladly add my promise to send dried or living specimens of the hybrids if the experiments should succeed.

With greatest respect for your honor, I sign myself,

G. MENDEL
Altbrünn, Monastery of St. Thomas

Brünn, 9 February 1868

V

HIGHLY ESTEEMED SIR:

Accept my most cordial gratitude for the *Hieracium* seeds, which arrived in good condition. How grateful I am for this kind shipment, and how much I do appreciate your kindness in promising also a shipment of living plants. I shall do my utmost to produce all the possible hybrids among the species, and if they should be fertile, their progeny will be observed for several generations. I must ask you to please charge the expenses of purchase and transportation, and any others, to my account.

I have received the news of your accident of March 1, esteemed Sir, with the greatest regret, and I am sincerely happy that the mishap was not accompanied by any worse consequences.

Recently there has been a completely unexpected turn in my affairs. On March 30 my unimportant self was elected life-long head, by the chapter of the monastery to which I belong. From the very modest position of teacher of experimental physics I thus find myself moved into a sphere in which much appears strange to me, and it will take some time and effort before I feel at home in it. This shall not prevent me from continuing the hybridization experiments of which I have become so fond; I even hope to be able to devote more time and attention to them, once I have become familiar with my new position.

On the whole the plants have wintered well in the experimental plots, and their development has progressed fairly well; most of the piloselloids and some of the *Archieracia* already show flower buds. The hybrids *H. Auricula* + *H. Pilosella, H. praealtum* (*Bauhini*) + *H. aurantiacum,* and probably also *H. Pilosella* + *H. Auricula* may be considered to have been successfully produced. About 100 of the autumn seedlings of last year's hybrid *H. praealtum* + *H. stoloniflorum* (Autor.) have survived the winter. Up to now these plants (still very small) are uniform in the structure and the hairy covering of the leaves, and resemble the hybrid seed plant. I am awaiting their further development with some suspense.

With the assurance of my deepest esteem I sign myself,

Your devoted friend,
GREGOR MENDEL
Abbot and Prelate of the Monastery of St. Thomas

Brünn, 4 May 1868

VI

HIGHLY ESTEEMED FRIEND:

Forgive me for being so tardy in expressing my most sincere gratitude for the species of *Hieracium* which you sent me. I received the little box on May 12. Since I had to start a long tour of inspection on the same day, I could not find the time necessary to thank you in writing. The gardener received instructions to handle the plants with great care, to pot one specimen of each, and plant the rest in the garden. When I returned a few days ago, I found to my great regret, that half of the potted plants had died, probably the consequence of excessive watering. The garden plants were well preserved, with few exceptions; they must be sorted, however, for the gardener neglected to add the names. The plants so lost include the piloselloids, with the exception of *H. flagellare, H. auriculaeforme,* and *aurantiacum,* and also *H. pulmonarioides* and *H. albidum.* I hope, however, that none of the species was lost in its entirety.

Of the seeds which you kindly sent me, the following have germinated: *H. amplexicaule, H. elongatum, H. alpinum, H. gothicum, H. tridentatum, H. praenanthoides. H. villosum, H. albidum,* and *H. glaucum* did not germinate.

The first generation of last year's hybrid *H. praealtum* + *H. flagellare*, consisting of 112 plants, is flowering. As far as I am able to judge, all plants are alike in the essential traits, and they differ from the hybrid seed plant, which is now flowering, only to the extent of having weaker, shorter, and less-branched stems. This is not remarkable in view of the greater age and strength of the seed plant. Fertility, as far as it can be judged at this time, is complete in all. The less favorable result obtained last year with the hybrid seed plant in this respect, might be explained by the fact that the latter was, while flowering, removed from the ground and transplanted to a pot; this must result in injury to the roots, and a weakening of the plant at the time of seed formation.

Five other piloselloid hybrids were obtained from last year's fertilizations:

1. *H. praealtum* (*Bauhini*) + *H. aurantiacum*. This hybrid is about intermediate between the parental species. The upper sides of the ray flowers are striped with orange, the lower sides striped with purple, the other flowers are golden or light yellow; the styles are yellow, the stigmas rusty brown. It was transplanted before starting to flower, therefore perhaps poorly fertile for that reason. I obtained another aberrant type from the same hybridization, but only two heads are opened thus far. In formation of leaves and stolons it resembles *H. praealtum* much more closely, and the stigmas are yellow; the stem, however, is bristly, and the flowers are definitely of hybrid color!

2. Another hybrid: *H. praealtum*(?) + *H. aurantiacum* has just started to flower. One of the parent plants appears to be intermediate between *H. praealtum* and *H. echioides*. The hybrid is of intermediate type, the flower color is as in the preceding plants; the stigmas are rusty brown.

3. *H. Auricula* + *H. Pilosella*, having the characteristic bifurcation of the stem, is also flowering. The heads are strikingly large, much exceeding the average, but this might be due only to the very luxuriant development of this plant.

4. *H. praealtum* (*Bauhini*) + *H. Pilosella*, and also *H.*

Auricula + *H. aurantiacum*, are just about to start flowering.

Plants from the spring sowing are not yet developed far enough to say anything about the success of the hybridization, but some success should have been obtained with them. They are mostly *Archieracia*.

Next spring my greatest concern will be to send you, as living specimens, all hybrids along with their respective parental species.

Your kind offer of supporting and augmenting my experimental flora in the future evokes my warmest gratitude, and inspires me, at the same time, to make the best possible use of the offered materials.

With the assurance of my deepest respect I sign myself,

Your devoted friend,
GREGOR MENDEL

Brünn, 12 June 1868

VII

HIGHLY ESTEEMED SIR AND FRIEND:

I am sending you as promised some hybrids of *Hieracium*, *Cirsium*, *Geum*, and *Linaria* which I have obtained by artificial fertilization. The catalogue is enclosed; the numbers correspond to those written on the labels tied to the plants.[3]

A. *Hieracia* (N.)

No. 1 was fairly fertile; the plants raised from seed thus far do not differ from each other morphologically. This hybrid, as well as Nos. 2, 5, 8, 9 are last year's and were raised in pots. (*H. calomastix* N., *H. magyaricum* + *H. aurantiacum* C.)

[3] The following enumeration did not appear in the original letter. It was written on the letter by Nägeli, presumably using a list which Mendel sent along with the specimens (vide Correns). Trs.

No. 2 was obtained from the same hybridization as No. 1, but is morphologically quite distinct and was quite infertile.

Nos. 3 and 4 are the two parental species. (H. Bauhini [later H. *magyaricum*], *H. auranticum* N.)

No. 5 bloomed prolifically, but produced only four viable seeds. (*H. Mendelii* N., *H. Auricula* + *H. Pilosella bruennense* C.)

Nos. 6 and 7 are the two parental species of the preceding hybrid. (*H. Auricula, H. bruennense* N., *H. Pilosella bruennense* C.)

No. 8 was completely sterile. The two parental species are Nos. 4 and 6. (*H. Auricula* + *H. aurantiacum* C.)

No. 9 was completely fertile. The progeny which are still young are uniform in their leaf formation. *H. monasteriale* N., *H. setigerum* + *H. aurantiacum* C.)

No. 10 is the parent of the preceding species. It grows on the wall of our monastery garden; thus far I know of no other location where it occurs. (*H. setigerum* N.) (The second parental species is No. 4.)

No. 11. Was only partially fertile when raised in a pot, but completely so when raised in the open ground. (*H. inops* N., *H. florentinum subcymigerum* + *H. flagellare* C.)

Nos. 12 and 13 are the two parental species. (*H. praealtum, H. flagellare* N.)

No. 14. All of last year's seedlings (there were 112) resembled the hybrid seed plant No. 11. All of them were fertile. (*H. inops* N., *H. flagellare* + (*florentinum*) *subcymigerum* C.)

B. *Cirsia* (N.)

No. 15. This is a vigorous, beautiful plant of medium fertility. Another irregular form was obtained from the same fertilization; it was sterile and died last summer.

Nos. 16 and 17 are the two parental types.

No. 18. I have already reported on the interesting progeny of hybrid No. 15 in my last letter. To my regret I must state today that I am only able to send four specimens, because the others, although they appeared to be vigorous and bloomed plentifully throughout the summer, died during the winter contrary to all expectations. I regret this the more, because, counting on a longer life span, I neglected to dry branches for the herbarium. It is hoped that the damage will be repaired, since the seed plant is healthy and vigorous. Considering the great hardiness of perennial *Cirsia*, it is remarkable that more than two thirds of the vigorous and luxuriant plants were lost after the first bloom. Were they destined by their composition to live only for two years? Or would they have lived longer under more favorable circumstances?

No. 19 is a very beautiful plant of average fertility. Although the seedlings are still young, their leaf formation leads me to expect as much variability as in the case of No. 18. Hybrids of *Hieracium* show, strangely enough, a very different behavior in the production of their progeny, than do those of *Cirsium*. *Cirsium* would be an excellent experimental plant for the study of variable hybrids, if it required less space.

Nos. 17 and 20 are the parental species of the preceding hybrid.

C. *Geum* (N.) *urbanum* + *rivale* (C.)

No. 21. Several hybrids from the same fertilization varied somewhat in flower size and were of unequal fertility.

Nos. 22 and 23 are the two parental types.

Nos. 24, 25, 26 will flower for the first time this year. Nos. 27 and 28 produced some incomplete flowers last year. According to Gärtner, progeny of this *Geum* hybrid show no variation.

D. *Linaria* (N.)

No. 29. With respect to flower color two different types of hybrids were obtained from the same fertilization; 33 hybrids had a rather yellowish, 21 a more purplish coloration, and one plant showed both colors. Fertility is low and the progeny show variation.

No. 31. A beautiful and vigorous hybrid, also of low fertility.

I have one specimen of the interesting hybrid *Mirabilis Jalappa* + *M. longiflora.* A few plants were obtained from the small number of seeds which it bore last summer; they are, however, still too delicate to stand transportation. The same is true of the hybrids:

H. cymosum + *H. Pilosella*
H. Auricula + *H. pratense*
(*H. praealtum* + *H. aurantiacum*) + *H. aurantiacum*
(*H. praealtum* + *H. aurantiacum*) + *H. praealtum*
(*H. Auricula* + *H. Pilosella*) + *H. Auricula*
Antirrhinum vulgare + *A. rupestre*
Lynchnis diurna + *L. vespertina*

which were produced during last year's fertilization experiments.

I have not yet succeeded in producing hybrids of *Archieracia,* but I hope that this year's sowing will yield results. Of the species which you so kindly sent me I was able to use only *H. humile, H. Sendtneri, H. picroides, H. prenanthoides, H. hispidum,* and *H. canescens* last summer. Since they, as well as the other species, have wintered well, I shall be able to extend my experiments. This shall be done exactly according to the plan which you, honored friend, were kind enough to send me.

For weeks we have been enjoying the most wonderful spring weather. As compared to several years' average, the vegetation is 13 days earlier than usual; almost everything has come into leaf already.

In gladly taking this opportunity to express to you, my most esteemed friend, my sincere respect and admiration and to commend myself to your continued benevolence, I sign myself,

Yours always respectfully,
GREGOR MENDEL

Brünn, 15 April 1869

VIII

HIGHLY ESTEEMED FRIEND:

Please do not be vexed with me for being so tardy in expressing my gratitude for the living *Hieracia*, all of which arrived safely, and are growing splendidly. Building operations on some remote dairy farms and other business matters have occupied me exclusively for many weeks, and when I returned to Brünn on Whitsunday, I found urgent and time-consuming tasks there also. I have been master of my own time for only a few days now, and in a position to resume my favorite occupation, which I had to discontinue about the end of June of last year, because of an eye ailment.

I found myself in serious danger of having to renounce my hybridization experiments completely, and this due to my own carelessness. Since diffuse daylight was not adequate for my work on the small *Hieracium* flowers, I had recourse to an illumination apparatus (mirror with convex lens), without suspecting what damage I might have done with it. After having occupied myself a good deal during May and June with *H. Auricula* and *H. praealtum*, a peculiar fatigue and exhaustion of the eyes appeared and reached a serious degree in spite of my immediately sparing my eyes as much as possible. It made me incapable of any exertion well into the winter. Since then the affliction has luckily been almost completely lost, so that I am again able to read for long stretches at a time, and can undertake the fertilization experiments in *Hieracium* as well as can be done without artificial illumination.

With today's letter I am sending some living *Hieracium* hybrids; where necessary the parental forms are also inclosed.

The results of the experiments thus far may be termed trifling, and are too incomplete to draw any final conclusions. Some experiences have been gathered, however, and I take the liberty to mention briefly what seems to me to be of some importance.

I must mention first that I have not yet succeeded in obtaining, by fertilization with foreign pollen, even a single hybrid in some species of piloselloids, in spite of numerous attempts. This is true, for instance, for *H. aurantiacum*. In this species I have not been able, thus far, to overcome the influence of its own pollen. *H. Pilosella* and *H. cymosum* also cause difficulties. In others, for instance, in the varieties of *H. praealtum*, fertilization with foreign pollen succeeds more easily with the same treatment, and I have repeatedly convinced myself that *H. Auricula* is a completely reliable experimental plant, if some care is employed. Last year I fertilized more than 100 heads of this species with pollen of *H. Pilosella, cymosum,* and *aurantiacum;* although about one half of them dried up because of injuries suffered, and only 2–6 seeds were obtained from each of the remaining plants, the plants raised from them are without exception hybrids. The tiny plants of *H. Auricula* + *H. Pilosella,* and *H. Auricula* + *H. cymosum* were, with few exceptions, unfortunately browsed upon by snails in the hothouse, but those of *H. Auricula* + *H. aurantiacum* were preserved; 98 of them have been planted in the garden. They should flower next month.

Still another type seems very suitable for experiments. I enclose it in the shipment under the designation No. XII, since I can neither name nor classify it. I have found it in large numbers on cut over lands. The one attempt at fertilization I made using it and the pollen of *H. Pilosella* was completely successful; all of the 29 plants obtained are hybrids.

I venture to state here that I have, thus far, used only a single form of *H. Pilosella* for fertilizations. Since, however, an adjacent type of *Pilosella* invaded the territory of my experimental plant, as I noticed later, at the time of flowering, I am not certain that a mistake was not made in last year's shipment, and I am enclosing the plant again, designated as *H. Pilosella* (Brünn). I do not dare to state an opinion as to whether this form stands in any kind of relationship to *H. echioides* or not, but will mention that it commonly occurs here, whereas the next known locality of *H. echioides* is about five miles away. The plant which I sent you under the designation of *H. praealtum*(?), was found there in association with *H. echioides* and *H. praealtum;* thus it is beyond doubt that the assumption that it belongs to the *H. echioides-praealtum* group was correct. A comparison with the parental species shows that it resembles *H. echioides* more closely.

I would be very grateful, honored friend, if you should give me, when convenient, your opinion on *Hieracium* No. XII. This plant, and *H. Auricula,* are among the best experimental plants, because a fairly large number of hybrids may rather easily be obtained from them. This condition is important, because the variations occurring among hybrid individuals can only be interpreted in cases where a fairly large number of hybrids is obtained from the same fertilization.

As a matter of fact, variants appeared in all those cases in which several hybrid specimens were obtained. I must admit to having been greatly surprised to observe that there could result diverse, even greatly different forms, from the influence of the pollen of one species upon the ovules of another species, especially since I had convinced myself, by growing the plants under observation, that the parental types, by self-fertilization, produce only constant progeny. In *Pisum* and other plant genera I had observed only uniform hybrids and therefore expected the same in *Hieracium*. I must admit to you, honored friend, how greatly I was deceived in this respect. Two specimens of the hybrid *H. Auricula* + *H. aurantiacum* first flowered two years ago. In one of them, the paternity of *H. aurantiacum* was evident at first sight; not so in the other one. Since, at the time I was of the opinion that there could be only one hybrid type produced by any two parental species, and since the plant had different leaves and a totally different yellow flower color, it was considered to be an accidental contamination, and was put aside. Thus, in last year's shipment I enclosed only the specimen which closely resembled *H. aurantiacum* in flower color. But when three specimens, each of the same hybrid produced from the fertilization in 1868, and also the hybrid *H. Auricula* + *H. pratense* (var.) later flowered, as three different variants, the correct circumstances could no longer escape recognition.

As I see from your treasured letter that the specimen of *H. Auricula* + *H. aurantiacum*, and the parental species *H. Auric.*, which I sent you, have died, I am replacing them, and also include the long misunderstood hybrid twin *H. Auricula* + *H. aurant.* 868b. Last year's three specimens are designated 869 c, d, e. The variant c is completely fertile.

During the winter, one variant of the hybrid *H. Auricula* + *H. pratense* (var.), and the parental species *H. pratense*, have died. The latter was not a typical *H. pratense*, since it carried some stellate hairs on the leaves. The two specimens made available to me through your kindness died during the first year in the garden; one became stunted without having flowered, the other during the flowering period. I have not yet been able to find the pure species in this vicinity.

The hybrid *H.* No. XII (*cymigerum* N.) + *H. Pilosella* (Brünn) has just started to wither. Very striking variations are seen among the 29 specimens available. Although they represent all transitional types from one parental species to the other, no one would take them for siblings if he found them growing in the wild. I shall send you the whole collection, as soon as the runners have rooted sufficiently; this should be the case in a few weeks. At that time I hope to be ready to report on this year's major experiment with *H. Auricula* + *H. aurantiacum;* because of the fairly large number of specimens I hope to get some information from it.

Hieracium No. XII has now been fertilized with *H. Pilosella vulgare* (München), and next year the comparison between the two hybrid series *H.* No. XII + *H. Pilosella* (Brünn) and *H.* No. XII + *H. Pilosella vulgare* (München) should not be without interest. *H. Auricula* has also been hybridized with *H. Pilosella vulgare* (M.) and *H. Pilosella* (Br.), shortly it will also be done with *H. Pilosella niveum* (M.). I have only seen one flower of *H. Pilosella incanum* as yet; it is hoped that others will appear.

Twenty-five plants of the hybrid *H. praealtum* (*Bauhini?*) + *H. aurantiacum,* of which I sent you two specimens last year, will come into flower soon. Differences between them exist, as far as may be observed at present. Of two specimens raised in pots, and already long past flowering, one is completely fertile, while the other is almost completely sterile. Both total sterility and complete fertility occur in the series of *H. Auricula* + *H. aurantiacum* hybrids.

The second generation of the hybrids *H. praealtum*(?) (*setigerum* N.) + *H. aurantiacum* and *H. praealtum* (*Bauhini?*) + *H. aurantiacum* has flowered, as has the third generation of *H. praealtum* + *H. flagellare.* Again the hybrids do not vary in these generations. On this occasion I can not resist remarking how striking it is that the hybrids of *Hieracium* show a behavior exactly opposite to those of *Pisum.* Evidently we are here dealing only with individual phenomena, that are the manifestation of a higher, more fundamental, law.

If one wants to follow the development of those hybrids having only partial fertility, it is necessary to protect the plants most carefully from the influence of foreign pollen, since individual ovules, which would normally remain unfertilized because of the predominantly poor quality of the pollen from the same plant, are readily fertilized by the pollen from other plants. I am sending at the same time some double hybrids, obtained from *H. praealtum* (*Bauhini?*) + *H. aurantiacum*; this plant being left to bloom and wither among plants of *H. Pilosella* (Brünn), but removed from other *Hieracia*. The hybrid should therefore be designated as (*H. praealtum* (*Bauh.?*) + *H. aurantiacum*) + *H. Pilosella* (Brünn). In more than one respect they are very interesting types.

If, in flowers of partially fertile hybrids, the stigmas are covered with the pollen of other, not too distantly related species, they always produce more seed than when kept isolated and dependent upon self-fertilization; that this is due exclusively to the action of the foreign pollen can easily be demonstrated by cultivation of the seeds. Careful isolation is, however, not necessary for completely fertile hybrids. Experiments with *H. praealtum*(?) + *H. aurantiacum* have shown that foreign pollen, even that of the two parental species, may be put upon the stigmas in quantity without interfering with self-fertilization. All seeds produce the original hybrid form.

I am adding the hybrid *H. cymosum* (München) + *H. Pilosella* (Brünn) to the shipment. It is the only hybrid of *H. cymosum* thus far obtained, although I have tried to fertilize this species many times.

In the *Archieracia* it is very difficult to prevent self-fertilization. Thus far only two hybrids have been obtained. The seed plant of one is that species with light brown seeds which I sent you once, as a dried specimen; the pollen was taken from a narrow-leaved *H. umbellatum*. The hybrid and the parental plants are enclosed. Among this year's seedlings, fertilization of a form of *H. vulgatum* with the above mentioned *H. umbellatum* can be recognized as having been successful. In vain I am looking for an *Archieracium* which might serve as well within its group as do *H. Auricula* and *Hieracium* No. XII among the piloselloids.

Of the *Archieracia* that I owe to your special kindness, all except *H. glaucum* have been subjected to experiments. They are the following: *H. amplexicaule, pulmonarioides, humile, villosum, elongatum, canescens, hispidum, Sendtneri, picroides, albidum, prenanthoides, tridentatum,* and *gothicum.* The artificially fertilized heads have always withered in *H. amplexicaule* and *H. albidum.* I do not have *H. alpinum.* From the seeds marked Breslau and München which you kindly sent me, *H. nigrescens* and one other species were obtained, that one, however, is not *H. alpinum.*

At this opportunity let me say that all my *Archieracia* are growing well. *H. albidum* is somewhat delicate when potted, especially during the winter, but it keeps well in the open ground. The same is true of the piloselloids, with the exception, however, of *H. pratense* and *H. Hoppeanum;* the latter died during the first winter in the open ground as well as in pots.

Because of my eye ailment I was not able to start any other hybridization experiments last year. But one experiment seemed to me to be so important that I could not bring myself to postpone it to some later date. It concerns the opinion of Naudin and Darwin that a single pollen grain does not suffice for fertilization of the ovule. I used *Mirabilis Jalappa* for an experimental plant, as Naudin had done; the result of my experiment is, however, completely different. From fertilizations with single pollen grains, I obtained 18 well-developed seeds, and from these an equal number of plants, of which 10 are already in bloom. The majority of the plants are just as vigorous as those derived from free self-fertilization. A few specimens are somewhat stunted thus far, but after the success of all the others, the cause must lie in the fact that not all pollen grains are equally capable of fertilization, and that furthermore, in the experiment mentioned; the competition of other pollen grains was excluded. When several are competing, we can probably assume that only the strongest ones succeed in effecting fertilization.

In fact, I want to repeat the experiment; and it should also be possible to prove directly by experiment whether or not two or more pollen grains can participate simultaneously in the fertilization of the ovule in *Mirabilis*. According to Naudin, at least three are needed!

Of the experiments of previous years, those dealing with *Matthiola annua* and *glabra*, *Zea*, and *Mirabilis* were concluded last year. Their hybrids behave exactly like those of *Pisum*. Darwin's statements concerning hybrids of the genera mentioned in "The Variation of Animals and Plants under Domestication," based on reports of others, need to be corrected in many respects.

Two experiments are still being continued. I have about 200 uniform specimens of the hybrid of *Lychnis diurna* and *L. vespertina*. The first generation should flower in August.

The color experiments with *Matthiola* have lasted now for six years, and will probably still go on for several years. With the data already obtained, I hope finally to get to the bottom of the problem. Lack of a reliable color chart has hindered the experiments greatly. Although I had ordered from Erfurt an assortment of *Matthiola annua* in 36 named colors, it proved insufficient for my purposes. I have given my special attention to this experiment, and I shall take the liberty to report on it as soon as the inspection of the 1500 specimens of this year's culture has been completed. It will be done at the same time as the shipment of the hybrid series *H.* No. XII + *H. Pilosella*.

In thanking you again, esteemed friend, for your kindness in making the shipment, I sign myself, with the expression of greatest respect,

Your devoted,

GR. MENDEL

Brünn, 3 July 1870

IX

HIGHLY ESTEEMED SIR AND FRIEND:

Together with this letter I am sending the 29 hybrids of *H.* No. XII (*H. cymigerum* + *H. Pilosella* (var. of Brünn) which I promised. The *Hieracium* designated No. XII (already sent you) is still an enigmatic form to me; might it perhaps be *H. poliotrichum* Wim.?

Eighty-four specimens of the hybrid *H. Auricula* + *H. aurantiacum* have flowered, some have died, others have not yet flowered. Variation among them was considerable. Each hybrid trait appears in a certain number of variants which represent different transitional stages between one ancestral trait and the other. It seems that the variants of the different traits may occur in all possible combinations. This seems probable because in the available hybrid plants the assortment of variants of the traits is exceedingly diverse, so as hardly to be the same in any two instances. If this assumption is correct, many hundreds of possible hybrid types should result because of the large number of differences between *H. Auricula* and *H. aurantiacum*. The observed number of hybrid types is too small in the case of parental species as distant as these to determine the true facts. Success should be obtained more easily with the hybrid *H. Auricula* + *H. Pilosella vulgare*; I hope to obtain about 200 specimens of it next year. *H. Auricula* + *H. Pilosella niveum* and *H. Auricula* + *H. Pilosella* var. of Brünn will be well represented. *H. Pilosella incanum* is completely sterile, and could not be fertilized with the pollen of the same *H. Auricula*.

I should like to mention that about one quarter of the hybrid types of *H. Auricula* + *H. aurantiacum* may be called completely fertile, one half partially fertile and one quarter sterile. The degree of fertility appears to be independent of the type of the hybrid.

If it should accord with your wishes, esteemed friend, I will send you the whole collection next spring.

Of this season's seedlings of *Archieracia* only a small portion has bloomed in this persistently cold and rainy weather; to date not a single specimen of *H. humile*, which I like to use as an experimental plant. The seedlings of the hybrid *H.? +H. umbellatum*, which I have already sent you, are not yet blooming either; they might still do so, however, if the autumn weather proves favorable. Thus far no differences among them are detectable. I would like to classify the doubtful parental species as *H. racemosum*, if the pale brown color of the seeds, shown thus far to be constant, were not present.

The experiment designed to solve the question whether or not a single pollen grain suffices for fertilization, was repeated with *Mirabilis Jalappa*, with the same results as last year. Plants obtained from last year's fertilizations using a single pollen grain can not be distinguished in any way from those produced by self-fertilization. In the beginning it seemed as if individual plants might lag behind in development; later, however, they completely made up the loss.

Under way is another experiment with *Mirabilis*, designed to find out also whether two pollen grains may simultaneously participate in fertilization. The varieties with crimson red, yellow, and white flowers, respectively, are constant when raised from seed, as I know from experience, and the hybrids which first result from the crosses *crimson + yellow* and *crimson + white* show no variations in their characteristic coloration. Both fertilizations succeed equally well and thus no differences in the degree of relationship [among the three varieties] is apparent. In the crimson variety a fairly large number of fertilizations was undertaken in such a way that two pollen grains were simultaneously put on each stigma, one of the yellow, and one of the white variety. Since the resultant flower colors of the crosses crimson + yellow and crimson + white are known, it will be shown next year whether in addition to the hybrid colors still a third color will appear, one explainable by joint action of the two pollen grains.

In the latter case, development of the progeny should also be different from that in the two simple color hybrids. These behave like *Pisum*, and half of the first generation again produces the hybrid color, while the other half receives the two parental colors in equal parts, and remains constant in the next generation. Those offspring of the hybrid crimson + yellow, which received the parental colors in the first generation, have also proved themselves to be constant as regards color in the second generation raised from seeds. Both colors reappear in pure form, as though they had never been in hybrid combination. Darwin and Virchow have pointed to the high degree of independence that is typical for individual characters and whole groups of characters in animals and plants. The behavior of plant hybrids indisputably furnishes an important proof of the correctness of this point of view.

The color experiments with *Matthiola annua* have made only minor progress this year in spite of the great number of experimental plants. According to experiences thus far, an agreement with *Pisum* appears probable. Certain phenomena concerned with the intensity of coloration cause difficulties. Instead of the expected color there frequently appears a higher or lower octave of color, if I may express it thus, or both appear jointly; and this happens not in isolated cases, but in a whole series of specimens. Thus sorting becomes very unreliable, because it is easy to put together what should be separated, or make the opposite error. The numbers thus obtained for the frequency of the different color variants are useless for the derivation of the series. Recently a new group of experimental plants has been studied; perhaps I will succeed in obtaining a simpler series with them.

Finally, let me report on a curiosity in the numerical ratios in which the male and the female plants of the hybrid *Lychnis diurna* + *L. vespertina* occur. I fertilized three flowers of *L. diurna* and planted the seeds of each capsule separately. They produced:

Capsule				
Capsule	I.	74 plants	54 female	20 male
"	II.	58 "	43 "	15 "
"	III.	71 "	54 "	17 "
		203	151	52

Is it chance only that the male plants occur here in the ratio 52:203 or 1:4, or has this ratio the same significance as in the first generation of hybrids with varying progeny? I should doubt the latter, because of the strange conclusions which would have to be drawn in this case. On the other hand the problem can not be so easily dismissed if one considers that the anlage for the functional development of either the pistil alone or of the anthers alone must have been expressed in the organization of the primordial cells from which the plants developed, and that this difference in the primordial cells could possibly be due to the ovules as well as the pollen cells being different as regards the sex anlage. Therefore I do not want to dismiss the matter completely.

In commending myself, esteemed Sir and friend, to your treasured consideration, I sign myself, with the expression of greatest respect,

Yours very devotedly,
GR. MENDEL

Brünn, 27 September, 1870

X

HIGHLY ESTEEMED SIR AND FRIEND:

Despite my best intentions I was unable to keep my promise given last spring. The *Hieracia* have withered again without my having been able to give them more than a few hurried visits. I am really unhappy about having to neglect my plants and my bees so completely. Since I have a little spare time at present, and since I do not know whether I shall have any next spring, I am sending you today some material from my last experiments in 1870 and 1871. If, because of the late season, the plants can no longer be put in the open ground, they will winter easily in a lighted area of an unheated greenhouse [Kalthaus], covered by moderately damp earth or sand.

I am sending you:[4]

1. *H. praealtum* (München) + *H. Pilosella incanum* (München)
2. *H. Auricula* + *H. Pilosella vulgare* (München)
3. *H. Auricula* + *H. Pilosella vulgare* (Brünn)
4. *H. Auricula* + *H. Pilosella niveum* (München)
5. *H. Auricula* + *H. aurantiacum* (Brünn)

In the following notes I must rely upon my records of 1871.

ad 1.	1	hybrid, completely sterile.
ad 2.	84	hybrids, all sterile, hardly any differences among them.
ad 3.	25	hybrids, all fertile, all uniform.
ad 4.	35	" all sterile, all uniform.
ad 5.	circa 90	" partially fertile, very variable.

The inflorescences in 1–4 are frequently simple, as in the parental species *H. Pilosella.*

The name following the hybridization symbol + refers in all cases to that species from which the pollen was taken. Thus the + sign has the meaning: fertilized with the pollen of . . . .

All the hybrids listed under each number were placed in isolated beds, one for each number, and thus did not disturb each other by growth of their stolons. All the hybrids of one group grew in the same bed, and intermingled to such an extent while they were without care and supervision, that sorting them is difficult, at times even impossible. For shipment I have selected only such specimens as I can assume with some certainty to have originated from separate

[4] [In the following list some synonyms have been omitted in this volume. *Eds.*]

hybrid seedlings. Only in No. 5 (*H. Auricula* + *H. aurantiacum*) was separation no longer possible, since extremely luxuriant specimens, which cover the bed like a carpet, occur in this hybrid. I am sending three completely fertile variants of this hybrid, which I had transplanted to a separate bed for a future experiment in the first summer after their seeds had matured. Two of them, which are closer to *H. aurantiacum*, have intermingled in the new location as well, and cannot be distinguished with certainty. Under No. 5, plants of both are found, I hope, in the envelope designated with *a*. The envelope *b* contains plants of the third variant, which resembles the parental species *H. Auricula* more closely.

Since the three hybrid variants just mentioned were shown to be completely fertile, they were to serve for studies of later generations, but these experiments were not executed. It can probably be assumed as rather certain that the progeny originating from self-fertilization of these variants will not be subject to the same variation shown in the original hybrids. At least the plants which were raised as a sample from those seeds formed by the variants, growing in the open ground without any protection, and among all the other hybrids, completely and without exception resemble the seed plant. *H. aurantiacum* was also flowering in the vicinity, and at the same time as the hybrid variants mentioned, without any influence of its pollen having been discernible.

Gärtner has proved the prepotency of the parental pollen over the pollen of hybrids in several species of plants. I have performed a single experiment with *Hieracium*, the result of which, although representing an isolated case, seems to deserve a brief report. The hybrid *H. praealtum* + *H. aurantiacum* was used as the experimental plant; it is only partially fertile, so that only one-fourth to one-third of the flowers of each head developed good seeds.

The experimental plant was raised in a pot at the window. After several heads had finished flowering, all those still flowering were removed, and only two of the heads still unopened were selected for the experiment. As soon as their first florets opened, the stigmas emerging from the anther tubes were immediately and thoroughly covered with pollen of *H. aurantiacum*. This was continued for three or four days until all florets had opened and all stigmas had been covered. At the time of maturation it became evident that each of the two artificially fertilized heads had formed considerably more seeds than had the other heads, which had been left to self-fertilization. The seeds of the latter were counted and the average number for each head was determined.

In the following year two types of plants were obtained from the seeds of the artificially fertilized heads: some completely resembling the hybrid seed plant, and others which were much closer to *H. aurantiacum*. Furthermore, numerical comparison showed that the number of those seedlings which had not deviated from the hybrid seed plant, and thus had originated by self-fertilization, was almost as great as it should have been, according to the determined average, if the two heads had been left to self-fertilization alone.

The pollen of *H. aurantiacum* is therefore effective only in those florets which would have remained sterile without interference, but it was unable to replace the hybrid pollen.

I want to emphasize here that I gave the utmost attention possible to this experiment which was, by the way, easily performed; that I never missed the hours of 7–9 in the morning, at which time daily a new zone of florets opens from the periphery of each flower disc toward the center; and that very fresh pollen of *H. aurantiacum* was transmitted to the stigma as soon as it made its appearance.

Far be it from me to interpret the result of this experiment as showing that Gärtner is wrong when he claims the pollen of the hybrid to be ineffectual in competition with the parental pollen. No proof to the contrary may be deduced from this experiment; the exception which *Hieracium* seems to make in this respect must find a natural explanation in the peculiar structure of the florets and the reaction of the organs of fertilization.

I suspect that free competition is excluded in this genus, as long as self-produced pollen is well developed and of good fertilizing ability, since then the foreign pollen would always come too late and could contend only unsuccessfully due to this fact. I have had frequent occasion to convince myself that, in the *Hieracia*, the anthers open within the bud, and transfer the pollen to the stigma, which they surround closely, so that the latter emerges from the tube already covered with pollen when the bud opens. In *H. aurantiacum*, *H. murorum*, and others, I have many times, at least a day before flowering, carefully severed the anther tube at its base and pulled it over the style without slitting it open on the side, and used all additional precautions possible, after which I covered the stigma repeatedly with the foreign pollen destined for the fertilization, and still have never raised anything except *H. aurantiacum* or *H. murorum* out of the obtained seeds.

On the basis of this experience I consider it likely that fertilization with foreign pollen can occur only if self-fertilization fails, as long as the ovule remains capable of fertilization; this seems not infrequently to be the case in this genus.

It is known that unfavorable changes in environmental conditions may result in reduced reproduction, therefore they may cause a sexual weakening or complete sterility, wherein the male organs always suffer first, as in animals in captivity. It should not be otherwise in plants. *H. Pilosella incanum* can not adapt itself very well to the local climate. In the summer the air here seems to be too dry, perhaps also too warm for this plant. In 1870 the May and June flowers were completely sterile, partially fertile in the following year, and toward autumn individual heads appeared to be completely fertile. Presumably the reason for this sterility was to be found, in the case of the summer flowers, in the poor quality of their own pollen, since I could not successfully fertilize *H. Auricula* with it either, while at the same time fertilization with the pollen of other *Pilosella* varieties caused no difficulties. Toward the end of August, however, fertilization with the pollen of *H. Pilosella incanum* was successful. Gärtner was also convinced by his experiments that the male principle (as he puts it) is always affected first.

If this were actually the case, the naturally-occurring hybridizations in *Hieracium* should be ascribed to temporary disturbances, which, if they were repeated often or became permanent, would finally result in the disappearances of the species involved, while one or another of the more happily organized progeny, better adapted to the prevailing telluric and cosmic conditions, might take up the struggle for existence successfully and continue it for a long stretch of time, until finally the same fate overtook it.

Species of which numerous hybrids have been shown to exist I would consider decrepit or would at least assume to be well past their prime (*H. Auricula, H. praealtum*).

I can not yet give a report on the success of the collection of Moravian hybrids of *Hieracium* initiated by Prof. Niessl. Shipments from the corresponding members of our society are expected not sooner than this winter.

With the expression of my greatest admiration and esteem I sign myself,

Yours very respectfully,
GR. MENDEL

Brünn, 18 November, 1873

Part Two 1881

References to Mendel's Work in W. O. Focke, Die Pflanzen-Mischlinge

Translated by the Editors

I. SYSTEMATIC INDEX OF KNOWN PLANT HYBRIDS

Pisum

Lit.: Th. A. Knight in Philos. Trans. 1799, II p. 195; Trans. Hort. Soc. London V p. 379; Gärtner Bast. S. 316; Darwin Variiren I Cap. 9, 11; Kreuz-u. Selbstbefr. S. 151; G. Mendel in Verh. naturf. Ver. Brünn IX Abh. p. 3ff. (*Page 108.*)

Knight and after him many other breeders have, by crossing peas, obtained numerous new strictly true-breeding seeds. It seems that the oldest varieties have lately begun to become weaker after 50 to 60 or more generations (Darwin, *loc. cit.*). Mendel's numerous crosses gave results quite similar to those of Knight, but Mendel thought that he found constant ratios between the hybrid types. In general, seeds produced by hybrid fertilization in peas, too, maintain the same shape and color that those of the mother plant had, even if these seeds produce plants that resemble the father plant completely and then also give rise to its kind of seed. On the other hand, examples are also cited in which the peas obtained through cross fertilization are said to have themselves shown the coloration of the pollen-contributing variety. (*Page 110.*)

Phaseolus

Ph. vulgaris L. var. *nanus* L. ♀ × *multiflorus* Lam. fl. *coccin.* ♂ has been produced artificially by G. Mendel. *Ph. nanus* is short, has white flowers and small white seeds, *Ph. multiflorus* is tall, winding, and has red flowers and colored (black and red) seeds. The hybrid plants, 17 specimens, in general resemble more the paternal parental strain *Ph. multiflorus*, yet the flowers are a paler red. Fecundity was rather low; from 17 plants 49 seeds were obtained, from which 31 flowering specimens were obtained in the following year. One of these had white flowers and white seeds, in the other the flower color ranged between red and pale violet; the seed color was just as variable. Fecundity was very uneven, but always inadequate, the red-flowering specimens being the least fecund on the average. Fecundity proved to be not inherited; on the contrary, the progeny of the most fecund specimens was sometimes completely sterile. (*Pages 111–112.*)

Hieracium

Lit.: G. Mendel in Verh. Naturf.-Ver. Brünn VIII. Abh. S. 26; Loret in Bull. soc. bot. Fr. VI p. 432; Fr. Schultz Arch. de flore p. 5, p. 61, p. 254; Rehmann in Oe. B. Z. XXIII p. 81ff.; Flora und florist. reports. (*Page 215.*)

The hybrids are polymorphic, according to Mendel's experiences, but the individual forms usually produce true-breeding seeds. (*Page 215.*)

Pilosella

HYBRIDS FROM *H. pilosella* L.

H. auricula L. × *pilosella* L. has been produced artificially by Fr. Schultz and G. Mendel. Schultz achieved reciprocal hybridization of the two species by transferring the pollen by means of a brush. He found *H. pilos.* ♀ × *auric.* ♂ identical with *H. auriculaeforme Fr.*, and *H. auric.* ♀ × *pilos.* ♂, occurring spontaneously near Bitsch, to be like *H. Schultesii,* whose leaves resemble more closely *H. auricula,* its flowers *H. pilosella.* Both hybrid forms showed luxuriant growth, but produced almost exclusively sterile seed. Mendel obtained only one specimen of his artificial *H. auric.* ♀ × *pilos.* ♂; this was somewhat fertile and produced constant offspring. (*Pages 215–216.*)

HYBRIDS OF *H. auricula* L.

G. Mendel produced artificially *H. auric.* ♀ × *prat.* ♂; he obtained 3 specimens, which differed considerably from one another and were moderately fertile; the progeny of each of these specimens closely resembled the mother plant. (*Page 218.*)

Mendel obtained *H. auricula* ♀ × *aurantiacum* ♂ in two considerably divergent specimens. One of these (*per–aurant.*) was sterile, the other (*per–auricula*) produced a single seed. (*Page 218.*)

HYBRIDS OF *H. aurantiacum* L.

H. praealtum Vill. ♀ × *aurantiacum* L. ♂ has been obtained by G. Mendel in two different moderately fertile specimens. The progeny of each resembled the mother plant. However, a specimen of the second generation reached fully normal fecundity. (*Page 218.*)

A single specimen of *H. echioides* Lumn. ♀ × *aurantiacum* L. ♂ was obtained by G. Mendel. It was fully fecund and true breeding, and did not produce reversions even after fertilization by pollen from the parental stock. (*Page 218.*)

HYBRIDS *of H. praealtum* VILL.

A single specimen of *H. praealtum* Vill. ♀ × *flagellare* Rchb. ♂ was obtained by G. Mendel. Its fecundity was nearly normal and its progeny constant. (*Pages 218-219.*)

II. HISTORY OF HYBRIDIZATION

[*From 1851 to the Present*]

Among the most recent scientific experiments on hybridization those of Rob. Caspary with *Nymphaeaceae*, of G. Mendel with *Phaseolus* and *Hieracium*, of D. A. Godron with *Datura*, *Aegilops* × *Triticum* and *Papaver* deserve to be called especially instructive. (*Page 444.*)

III. ORIGIN OF HYBRIDS

The experiments of Kölreuter, Wiegmann, Gärtner, Godron, Naudin, Wichura, Mendel, Caspary, and others served only scientific purposes, while Herbert and Regel combined scientific and horticultural ones. (*Page 459.*)

IV. PROPERTIES OF HYBRIDS

The different primary forms of *Hieracium* hybrids were found by Mendel to be true breeding. (*Page 483.*)

V. NOMENCLATURE OF HYBRIDS

It did not occur to any one of the scientific hybridizers to assign special species names to their newly created plant forms. In this matter Kölreuter and Gärtner, Wiegmann and Lehmann, Naudin and Godron, Wichura, Mendel, and Caspary have acted alike. (*Page 492.*)

Part Three 1900

The Law of Segregation of Hybrids

Das Spaltungsgesetz der Bastarde (Preliminary Communication) [1]

HUGO DE VRIES

Submitted March 14, 1900

Translated by Evelyn Stern

According to pangenesis the total character of a plant is built up of distinct units. These so-called elements of the species, or its elementary characters, are conceived of as tied to bearers of matter, a special form of material bearer corresponding to each individual character.[2] Like chemical molecules, these elements have no transitional stages between them.

For many years this principle has represented the starting point for my investigations. Many important consequences can be deduced from it and may be tested experimentally. My experiments lie in part in the realm of variability[3] and mutability and in part in that of hybridization.

[The original paper was published in Berichte der deutschen botanischen Gesellschaft 18 (1900): 83–90.]

[1] Detailed description of my experiments and the theoretical analysis I intend to publish in a rather extensive work on the empirical units of species traits and their origin: "Mutation Theory."

[2] [Hugo de Vries] Intracelluläre Pangenesis [1889] pp. 60–75. For the opposite point of view, that every bearer of matter represents the total species character, compare pp. 47–60 of the above.

[3] Berichte der deutschen botanischen Gesellschaft 12 (1894): 197.

In this latter area, however, a complete change in the viewpoint from which the investigation proceeds is necessary. The altered viewpoint requires that "*the concept of species recede into the background in favor of the consideration of a species as a composite of independent factors.*"[4]

Current doctrine regarding hybrids considers the species, the subspecies, and the varieties as the units whose combinations create hybrids and which should be studied. One distinguishes between mixtures of varieties and the true species hybrids. Depending on the number of parental types one speaks of diphyletic or polyphyletic hybrids, of triple or quadruple hybrids, and so on.

In my opinion this way of looking at the problem should be abandoned in physiological investigation, for although it suffices for systematic and horticultural purposes, it is inadequate for the purpose of obtaining more basic knowledge of species.

The *principle of the crossing of species-specific traits* should replace it. The units of species-specific traits are to be seen in this connection as sharply separate entities and should be studied as such. They should be treated as independent of each other everywhere, as long as there is no basis for doing otherwise. In every crossing experiment only a single character or a definite number of them is to be taken into consideration: the others can be disregarded temporarily. Or rather, it is a matter of indifference whether the parents are distinguishable from each other in still further ways. However, for experimental purposes the simplest conditions are presented by hybrids whose parents differ from each other in one trait only (*monohybrids,* in contrast to the *di-* to *polyhybrids*).

If the parents of a hybrid differ from each other in one point only, or if only one or a few of their points of difference are selected for consideration, in these characteristics

[4] Intracelluläre Pangenesis, p. 25.

Hugo de Vries, 1896. After an etching by Jan Veth.

they are *antagonistic,* while in all other respects they are alike or indifferent for the analysis. The crossing experiment is thereby limited to the antagonistic characteristics.

My experiments have led me to make the two following statements:[5]

1. *Of the two antagonistic characteristics, the hybrid carries only one,* and that in complete development. Thus in this respect the hybrid is indistinguishable from one of the two parents. There are no transitional forms.

2. *In the formation of pollen and ovules the two antagonistic characteristics separate,* following for the most part simple laws of probability.

These two statements, in their most essential points, were drawn up long ago by Mendel for a special case (peas).[6] These formulations have been forgotten and their significance misunderstood.[7] As my experiments show, they possess generalized validity for true hybrids.

The lack of transitional forms between any two simple antagonistic characters in the hybrid is perhaps the best proof that such characters are well delimited units.[8]

And to support the correctness of this statement innumerable cases can be advanced, partly from my own experience, and partly from the literature. The fact that polyhybrids so frequently represent intermediate forms obviously depends on the fact that they inherited one part of their traits from the father and the other part from the mother. In monohybrids such is not possible.

[5] The "false hybrids" of Millardet are temporarily disregarded altogether in what follows.

[6] Gregor Mendel, Versuche über Pflanzen-Hybriden, in Verhandlungen des naturforschenden Vereines in Brünn 4 (1865): 3. This important treatise is so seldom cited that I first learned of its existence after I had completed the majority of my experiments and had deduced from them the statements communicated in the text.

[7] See G. and A. Focke, Die Pflanzenmischlinge, p. 110. [Reference is to W. O. Focke, Die Pflanzen-Mischlinge. Eds.]

[8] Intracelluläre Pangenesis.

Of the two antagonistic characters, Mendel calls the one visible in the hybrid the *dominating,* the latent one *recessive.*

Ordinarily the character higher in the systematic order is the dominating one, or, in cases of known ancestry, it is the older one. For example,

dominating	recessive
Papaver somniferum (tall form)	*P. s. nanum*
Antirrhinum majus, red	*A. m. album.*
Polemonium coeruleum, blue	*P. c. album.*

And where the ancestral background is known, for example:

dominating	recessive	known since
Chelidonium majus	*C. laciniatum*	±1590
Oenothera Lamarckiana	*O. brevistylis*	±1880
Lychnis vespertina (hairy)	*L. v. glabra*	±1880

Using this rule analogously in other cases, at times one arrives at contradictions of the prevailing systematic interpretation, for example:

dominating	recessive
Datura Tatula	*D. Stramonium*
Zea Mays (naked seed)	*Z. cryptosperma*

In species hybrids (polyhybrids), where the relative ages of the parental forms are usually unknown, possibly conclusions may be drawn from crossing experiments, for example, with reference to flower color:

dominating	recessive
Lychnis diurna (red)	*L. vespertina* (white)

THE LAW OF SEGREGATION OF HYBRIDS

In the hybrid the two antagonistic characters lie next to each other as anlagen. In vegetative life only the dominating one is usually visible. Exceptions occur seldom; an example is presented by some sectional segregations. Thus *Veronica longifolia* (blue) × *V. longifolia alba* in my experiments not infrequently forms racemes whose flowers are white on one side and blue on the other.

In the formation of pollen grains and ovules these characters separate. The individual pairs of antagonistic characters behave independently during this process. From this separation the law can be deduced:

The pollen grains and ovules of monohybrids are not hybrids but belong exclusively to one or the other of the two parental types. The same holds for di- to polyhybrids with reference to each character taken by itself.[9]

The composition of the progeny can be calculated from this statement, and by means of this calculation the validity of the statement can be proven experimentally. In the simplest case segregation will take place obviously into two equal halves and so one obtains:

50% dom. + 50% rec. pollen grains, and
50% dom. + 50% rec. ovules.

If dominating is designated by d and recessive by r, fertilization yields:

$$(d + r)(d + r) = d^2 + 2\,dr + r^2$$

or

$$25\%\, d + 50\%\, dr + 25\%\, r.$$

The individuals d and d^2 have only the dominating character, those of r and r^2 constitution possess only the recessive character, while the dr plants are obviously hybrids.

In self-fertilization, whether this takes place in isolation or in groups, the hybrids of the first generation yield, with reference to each single trait,

[9] The combinations take place according to probability calculus.

25% of individuals with the paternal trait,
25% " " " " maternal " ,
50% " " that are again hybrids.

According to our first main statement the hybrids possess the dominating trait, so that one obtains

75% of individuals with the dominating trait,
25% " " " " recessive " .

I found this composition confirmed in many experiments. For example:

A. *Following artificial crossing:*

Dominating	Recessive	Rec.	Year of Crossing
Agrostemma Githago	*nicaeensis*	24%	1898
Chelidonium majus	*laciniatum*	26%	1898
Hyoscyamus niger	*pallidus*	26%	1898
Lychnis diurna	*L. vespert.* (white)	27%	1892
" *vespertina* (hairy)	*glabra*	28%	1892
Oenothera Lamarckiana	*brevistylis*	22%	1898
Papaver somnif. Mephisto	*Danebrog*	28%	1893
" " *nanum* (simple)	filled	24%	1894
Zea Mays (starchy)	*saccharata*	25%	1898

B. *Following free crossing, for example:*

Dominating	Recessive	Rec.	Year of Crossing
Aster Tripolium	*album*	27%	1897
Chrysanthemum Roxburghi (yellow)	*album*	23%	1896
Coreopsis tinctoria	*brunnea*	25%	1896
Solanum nigrum	*chlorocarpum*	24%	1894
Veronica longifolia	*alba*	22%	1894
Viola cornuta	*alba*	23%	1899

The mean of all these experiments is 24.93%.

The experiments usually included several hundred plants and at times about a thousand. I obtained corresponding results with many other species.

Distinguishing the remaining 75% into the two groups listed is much more troublesome. This requires that a number of plants bearing the dominating trait be fertilized with their own pollen and that in the succeeding year the progeny be cultivated and counted for each plant. I carried out this experiment in 1896 with *Papaver somniferum* Mephisto × Danebrog and obtained accordingly for the composition of the first generation of 1895 the following:

Dominating (Mephisto)	24%
Hybrids (with ±25% Danebrog) .	51%
Recessive (Danebrog)	25%

This result is concordant with the formula cited above, or more correctly expressed, it was from these numerical relations that I first deduced the formula.

The dominating and the recessive traits are shown to be constant in the progeny, so far as they were isolated by segregation. However, the hybrids segregated again according to the same law. In this experiment they yielded an average of 77% with the dominating and 23% with the recessive trait.

Such behavior remains unchanged during the course of years. I extended this experiment through two further generations. The 50% of hybrids segregated out while the 25% with dominating trait remained constant.

From the main statement of the law of segregation various other inferences may be drawn, by means of which experimental testing becomes possible.

For example, by fertilizing a hybrid with the pollen of one of the two parents, or, in reverse, fertilizing one of the parental types with the hybrid, one obtains:

$$(d + r)d = d^2 + dr$$

and

$$(d + r)r = dr + r^2.$$

Thus in the first case some plants that develop are hybrids, some pure forms, but all of them exhibit the dominating trait. In the second case there are equal numbers of hybrids with the dominating trait and of pure specimens, so that one sees:

50% dominating (hybrids)
50% recessive (pure).

I found, for instance:

	Recessive	Year
Clarkia pulchella ×× white	50%	1896
Oenothera Lamarckiana ×× *brevistylis*	55%	1895
Silene Armeria (red) ×× white	50%	1895

The same law holds also, as already mentioned, if one investigates dihybrids or studies two pairs of antagonistic traits in polyhybrids. I choose as an example a cross of the thorny *Datura Tatula* with *Datura Stramonium inermis* which I made in 1897. In accord with a well-known rule the hybrids are all alike, regardless of which form contributed the ovules and which the pollen. They bloom blue and bear prickly fruit. Some flowers were fertilized by their own pollen and the resulting seed sown in 1899. By the time of germination the blue-blooming plants were distinguishable from the white-blooming ones by the color of the stems. I found:

Blue (domin. + hybr.) 72%
White (recessive) 28%

as was confirmed by the flower. With reference to the fruit there were produced:

Thornless, among the blues	26.8%
" " " whites	28.0%
Mean . . .	27.4%

From this the composition of the progeny can be calculated for nearly all cases. If, for instance, one pair of antagonistic characters is called A and the other pair B, one obtains for dihybrids:

A. 25% Dom.

B. 6.25 *d*, 12.5 *dr*, 6.25 *r*

A. 50% D × R   25% Rec.

B. 12.5 *d*, 25 *dr*, 12.5 *r*   6.25 *d*, 12.5 *dr*, 6.25 *r*

Thus 6.25% of the cases are pure dominating in both respects, and an equal number pure recessive in both, etc.

If one applies the rule that hybrids exhibit the dominating trait, one finds for the visible characteristics of the progeny:

1. A. dom. + B. rec.	18.75%
2. A. rec. + B. dom.	18.75%
3. A. dom. + B. dom.	56.25%
4. A. rec. + B. rec.	6.25%

As evidence I cite the following experiment. *Trifolium pratense album* was crossed with *Trifolium pratense quinquefolium;* the white flowers and the ternate leaves are recessive to the antagonistic species traits. I found for the progeny of the hybrids:

		Calculated
1. Red and ternate	13%	19%
2. White and pentad	21%	19%
3. Red and pentad	61%	56%
4. White and ternate	5%	6%

in approximately 220 plants.

In similar fashion calculations and experiments are to be applied to tri- to polyhybrids.

Success is frequently had in separating simple characters into a number of factors by means of segregation. For example, the color of flowers is often composite, and after crossing one obtains the individual factors, partly separate and partly in various mixtures. I have carried out such analyses with *Antirrhinum majus*, *Silene Armeria*, and *Brunella vulgaris* and found in them the above numerical relationships confirmed. *Antirrhinum majus* red, for example, may be split by crossing with white into both these colors and into yellow with red (Brilliant) and white with red (Delila); *Silene Armeria* may be split into red, pink, and white. *Brunella vulgaris* forms a constant intermediate form with a white flower and a brown calyx.

From these and numerous other experiments I draw the conclusion that the law of segregation of hybrids as discovered by Mendel for peas finds very general application in the plant kingdom and that it has a basic significance for the study of the units of which the species character is composed.

Carl Correns, 1905.

G. Mendel's Law Concerning the Behavior of Progeny of Varietal Hybrids

G. Mendel's Regel über das Verhalten
der Nachkommenshaft der Rassenbastarde

CARL CORRENS
Submitted April 26, 1900

Translated by Leonie Kellen Piternick

The latest publication of Hugo de Vries: "Sur la loi de disjonction des hybrides," [1] which through the courtesy of the author reached me yesterday, prompts me to make the following statement:

In my hybridization experiments with varieties of maize and peas, I have come to the same results as de Vries, who experimented with varieties of many different kinds of plants, among them two varieties of maize. When I discovered the regularity of the phenomena, and the explanation

Berichte der deutschen botanischen Gesellschaft 18 (1900): 158–168. Reprinted in Carl Correns, Gesammelte Abhandlungen zur Vererbungswissenschaft aus periodischen Schriften 1899–1924 (Fritz von Wettstein, ed.) Berlin, Julius Springer, 1924, pp. 9–16. [The translation printed here was originally published in Genetics 35, no. 5, pt. 2 (1950): 33–41; with the approval of the translator minor changes have been made in the text. *Eds.*]

[1] Compt. rend. de l'Acad. des Sciences (Paris), 130 (1900) 26 mars. [The paper referred to is a shorter version, in French, of the German paper printed in English translation on pp. 107–117 of this volume. *Eds.*]

thereof—to which I shall return presently—the same thing happened to me which now seems to be happening to de Vries: I thought that I had found *something new.*[2] *But then I convinced myself that the Abbot Gregor Mendel in Brünn, had, during the sixties, not only obtained the same result through extensive experiments with peas, which lasted for many years, as did de Vries and I, but had also given exactly the same explanation, as far as that was possible in 1866.*[3] Today one has only to substitute "egg cell" or "egg nucleus" for "germinal cell" or "germinal vesicle" and perhaps "generative nucleus" for "pollen cell." An identical result was obtained by Mendel in several experiments with *Phaseolus*, and thus he suspected that the rules found might be applicable in many cases.

Mendel's paper, which although mentioned, is not properly appreciated in Focke's *Die Pflanzen-Mischlinge*, and which otherwise has hardly been noticed, is among the best that have ever been written about hybrids, in spite of some objections which one might raise with respect to matters of secondary importance, e.g., terminology.

At the time I did not consider it necessary to establish my priority for this "re-discovery" by a preliminary note, but rather decided to continue the experiments further.

In the following I shall limit myself to an account of the experiments with varieties of peas.[4] Intervarietal hybrids of maize show identical behavior in all essential points, but are more difficult to experiment with, and I have not yet elucidated to my satisfaction several points of secondary importance. They will be discussed elsewhere in more detail.

Varieties of peas are invaluable for the problem which interests us here, as Mendel emphasized correctly, since the

[2] See the postscript. (Footnote added later.)

[3] Gregor Mendel, Versuche über Pflanzen-Hybriden, Verhandlungen des naturforschenden Vereines in Brünn 4 (1865), Abhandlungen, pp. 3–47.

[4] The names of the varieties given in this paper are those which I received from [the seed firm] Haage and Schmidt in Erfurt.

flowers are not only autogamous, but are very rarely fertilized by insects. On the basis of experiments on the formation of xenia—which, in the case of peas yielded only negative results—I came across this material. When I realized that the rules here are much clearer than they are in maize, where I had first discovered them, I continued the observations.

The *traits* which differentiate the varieties of peas, can, as in all other cases, be grouped into *pairs*, each member having an effect on the same trait, one in one and the other in the other one of the varieties: e.g., the color of the cotyledons, the color of the flower, the color of the seed coat, the hilum of the seed, etc. In *many* pairs one trait, or rather the anlage thereof, is so much stronger than the other trait, or its anlage, that the former alone appears in the hybrid plant, while the latter does not show up at all. This one may be called the *dominating*, the other one the *recessive*, anlage. Mendel named them in this way, and, by a strange coincidence, de Vries now does likewise. For example, the yellow color of the cotyledons dominates over the green color, and red flower color over white flower color.

I can not understand why de Vries assumes that in *all* pairs of traits which differentiate two strains, one member must always dominate.[5] Even in peas, where some traits completely conform to this rule, other trait pairs are also known in which neither trait dominates, as for instance the color of the seed coat being either reddish-orange or greenish-hyaline.[6] In this case the hybrid may show all transitions, (this is true especially for the seed coat of peas), or may show either more of one or of the other trait (for example

[5] For instance "D'autre part, l'étude des caractères simples des hybrides peut fournir la preuve la plus directe du principle énoncé. *L'hybride montre toujours le caractère d'un des deux parents, et cela dans toute sa force;* jamais le caractère d'un parent, manquant à l'autre, ne se trouve réduit de moitié." (*loc. cit.* paragraph 3, italics mine).

[6] The color of the hilum on the other hand (whether black, brownish etc.) represents a dominating character.

in hybrids of stocks; here a certain hybrid may be *just barely* distinguished from one parental form by its hardly noticeable, slighter covering of hairs, although with some care separation is always possible, while it is *highly distinct* from the other, i.e., the glabrous parental type).

The following holds only for pairs of traits which have a dominating and a recessive member; there is no reason to believe that it may not hold for other types of pairs of traits as well, but at present we know of no example.[7] Let us first consider a *single* pair of traits. It is immaterial whether the varieties to be crossed are differentiated only by this one pair, or by others as well. The specific pair we may select is the *color of the embryo, either yellow or green*. It is very easy to obtain large numbers for this trait.

The facts, which Mendel found, I can fully confirm. They also agree with the findings of de Vries for his experimental objects. They are as follows:

1. In the *first* generation, all hybrid individuals are uniform and only the *dominating* character appears. In our special case the cotyledons are *yellow*.

2. When these seeds with yellow embryos are sown, plants are obtained whose pods, which were produced by self-fertilization, contain seeds with *yellow* embryos and seeds with *green* embryos (the *second* generation), and on the average, there are *three* yellow ones for each green *one*. If there are four or more seeds in each pod, one containing a green embryo will usually be among them.

3. When the seeds with a *green* embryo are sown, plants are obtained whose pods, which were produced by self-fertilization, contain *only* seeds with *green* embryos (the *third* generation). These, in turn, produce only seeds with *green* embryos (the *fourth* generation),

[7] In the meantime I have found an example. (Footnote added later.)

etc. With respect to this trait, the *recessive* one, they behave like the *pure* variety, which carries it.

4. If the seeds with *yellow* embryo are sown, plants are produced which may be grouped into *two classes,*

Class A, those plants whose pods, which were obtained by self-fertilization, contain *only* seeds with *yellow* embryos (the *third* generation) and

Class B, those plants whose pods, which were produced by self-fertilization, contain seeds with *yellow* as well as seeds with *green* embryos (the *third* generation). Numerically, there are again on the average *three* seeds with *yellow* embryos for each *one* with a *green* embryo, just as in the second generation (see paragraph 2).

The *number of individuals* in classes A and B is approximately *one* to *two.*

Let me emphasize again, that *embryos* of Class A do not differ in their *appearance* in any way from those in Class B; only after the pods which were produced by self-fertilization have been harvested can it be decided to which one of the classes the seed belonged.

5. Seeds with *yellow* embryos, which descended from plants of *Class* A (paragraph 4), produce plants whose pods, which originated by self-fertilization, again contain *only* seeds with *yellow* embryos (the *fourth* generation). Plants which develop from them in turn produce *only* seeds with *yellow embryos* (the *fifth* generation), etc. As regards this trait, the *dominating* one, they behave like the *pure* variety which carries it.

6. The seeds with *green* embryos, which are obtained from plants of *Class B* (paragraph 4, B), produce plants whose pods, which originated by self-fertilization, again contain only seeds with *green* embryos (the *fourth* generation). Plants which develop from them in turn produce *only* seeds with *green* embryos (the *fifth* generation), etc.—just as did the green embryos of the *second* generation (paragraph 3).

TABLE I

Parents	Hybrid					
	I. Gen.	II. Gen.	III. Gen.	IV. Gen.	V. Gen.	VI. Gen.
∞ green		1 green	∞ green	∞ green	∞ green	∞ green
			1 green	∞ green	∞ green	∞ green
				1 green	∞ green	∞ green
					1 green	∞ green
						1 green
					2 yellow	
	∞ yellow	2 yellow	2 yellow	2 yellow	3	3 yellow
					1 yellow	
				3		
						∞ yellow
		3	3			
				1 yellow	∞ yellow	∞ yellow
			1 yellow	∞ yellow	∞ yellow	∞ yellow
∞ yellow		1 yellow	∞ yellow	∞ yellow	∞ yellow	∞ yellow

7. The seeds with *yellow* embryos, which are obtained from plants of *Class B* (paragraph 4, B) again produce, just as described in paragraph 4, *two types of plants,* in the ratio *one* to *two,* whose seeds behave in the same way as described in paragraphs 5 and 6 and so forth.

Table 1 explains and summarizes the results discussed above; it also gives the numerical ratios. (The sign ∞ indicates that all of the seeds of the progeny in this group contained like embryos.)

The two following tables show the results obtained in two of my experimental series. The generations are given in vertical sequence. The upper *figure in bold face* denotes in each generation the *number of embryos obtained,* the *figure in light face* the *number of individuals,* which were raised from these embryos and produced fruits; ye = yellow, gr = green. The rest is self-explanatory.

The numerical ratios of yellow embryos to green embryos are quite variable in individual plants. In experiment I the smallest percentages for green are 14.9 and 7.7 and the largest ones 44.2 and 40.0. It is of no importance whether the dominating trait was introduced by the paternal or by the maternal plant; this is true of *all* varieties that possess a specific pair of traits.

Experiment II shows, by chance, the exact numerical ratios between the two classes of individuals produced by seeds with yellow embryos (7:14 = 1:2), while this ratio can be determined only from the mean of generations III and IV in Experiment I: 15 [= 7(III) + 8(IV)] individuals in one class as opposed to 28 [= 18(III) + 10(IV)] of the other (34.9 : 65.1 instead of 33.3 : 66.6).

In order to *explain* the facts, one must assume (as did Mendel) that following fusion of the reproductive nuclei[8] the "anlage" for one trait, the "recessive" one (*green* in our

[8] Mendel, of course, does not mention nuclei, but only "germinal cells" and "pollen cells."

case), is suppressed by the other trait, the "dominating" one; therefore all embryos are *yellow*. The anlage, however, although "latent" is preserved, and prior to *the definitive formation of the reproductive nuclei a complete separation of the two anlagen occurs, so that one half of the reproductive nuclei receive the anlage for the recessive trait, i.e., green, the other half the anlage for the dominating trait, i.e., yellow.*

TABLE II

Experiment I. Hybrid between the "green late [variety] Erfurter Folgererbse" *with* green *embryos and the* "[variety] Kneifelerbse with purple-violet pods" *with* yellow *embryos*[9]

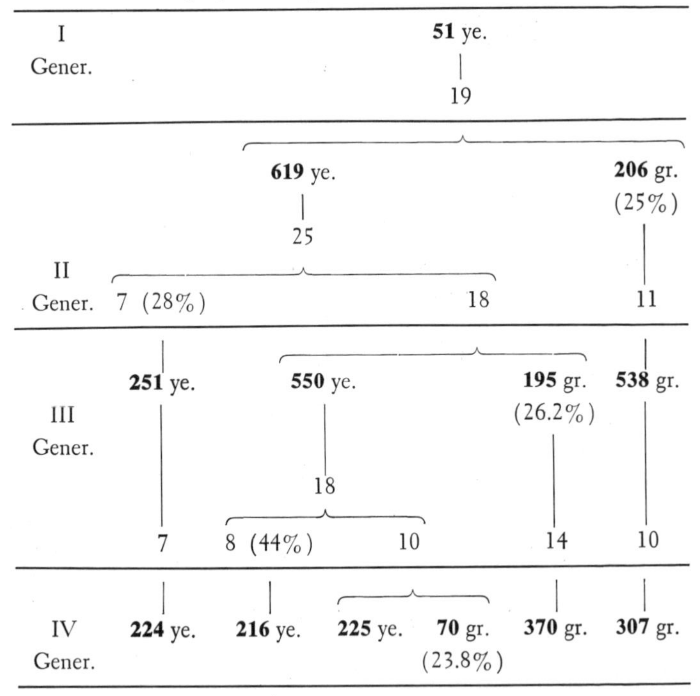

[9] Under identical conditions the plants produced an average of 43.3, 47.7, and 28.8 seeds in successive generations; this is a good example of the consequences of self-fertilization, and also furnishes an explanation of the "giant growth" of some hybrids. (Footnote added later.)

The earliest time at which this separation might occur is the time of formation of the primordial anlage of both the seed and the anthers.[10] The numerical ratio 1:1 strongly suggests that the separation occurs during a *nuclear division,* the reduction division of Weismann,[11] but, because of the numerous problems involved, a more detailed discussion would lead too far.

TABLE III

Experiment II. Hybrid between the "green late [variety] Erfurter Folgererbse," *with* green *embryos and the* [variety] "Bohnenerbse," *with* yellow *embryos*

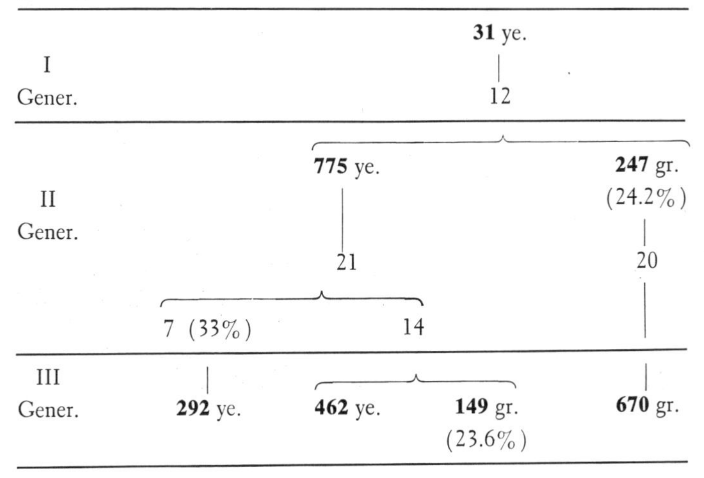

[10] At the latest at the time of the first division of the pollen grain nucleus from which the primary nucleus of the embryo sac is formed. For, in maize, it is shown by the similarity between the hybrid endosperm and the hybrid embryo that the two generative pollen tube nuclei and all of the eight nuclei of the embryo sac contain only one of the two anlagen. (Footnote added later).

[11] See also "Keimplasma," p. 392ff.

Thus among 1000 ovules, 500 contain the anlage for the dominating trait (yellow), 500 the anlage for the recessive trait (green), and among 1000 generative pollen tube nuclei there are also 500 each with the dominating (yellow) and with the recessive trait (green). If the reproductive nuclei are brought together by chance, then the probability that among 1000 nuclear fusions two anlagen *of the same kind* will meet (either two dominating or two recessive ones) is equal to the probability of *two different* anlagen meeting (one dominating, one recessive). Thus each type of combination will occur 500 times, or in 50 percent of all of the combinations.

In the first case, i.e., when *like* anlagen meet, the probability that they will be two *recessive* ones is as great as that they will be two *dominating* ones, again one half; each one will occur 250 times or in 25 percent of all combinations. For the pair of traits under investigation, the result is here the same as it would be if two reproductive nuclei of either one of the pure varieties should unite.

In the second case, when *different* anlagen are combined, the result of self-fertilization must be the same as that found in hybrids of the first generation, which were produced by experiment. The dominating anlage suppresses the recessive one, but later, preceding definitive formation of the reproductive nuclei, the two anlagen separate again, as was described for the artificially produced hybrid. "Thus, repeated hybridization takes place." (Mendel).

The progeny of the first generation must consequently be separable into three classes, 25 percent having *only* the recessive, 25 percent having *only* the dominating, and 50 percent having *both* traits, although [in the latter] only the dominating trait may be recognized. It follows from this assumption that in the first two cases all future generations will breed true for one of the two traits, while in the third case segregation will occur again.

If the hybrid (first generation) is pollinated with pollen

of that parental variety which has the *dominating* trait, instead of with its *own* pollen, only plants which show the *dominating* trait are obtained, but among their progeny *one half* will in turn produce *only* individuals with the *dominating* trait, while *the other half* will produce some plants with the *dominating* and others with the *recessive trait,* in a ratio of 3:1. If, on the other hand, the hybrid (first generation) is fertilized with the pollen of the parental variety with the *recessive* trait, then *one half* of the plants obtained will show the *recessive* trait, while the *other half* will show the *dominating* trait, and the progeny of the *latter* will again show the *dominating and* the *recessive* traits in a ratio of 3:1.

This theoretically derived rule also holds in the hybrids of maize.

Since two classes of individuals, i.e., those with the dominating anlage only and those with both the dominating and the recessive anlagen, cannot be distinguished from one another externally, the correct numerical ratios can be determined only by *self-fertilization.* Since self-fertilization normally occurs in peas, they are excellent experimental objects.

A further consequence of the above is the following: as long as, because of chance selection, the number of individuals in a plot remains constant in successive generations, the number of individuals in the modal class, i.e., those containing both anlagen, decreases steadily, until they finally disappear completely. In the second generation they make up 50 percent of the total, 25 percent in the third generation, 12.5 in the fourth, 6.25 in the fifth, and $100/2^{n-1}$ percent in the *n*th generation. This numerical decrease of the modal class had been derived by Mendel.[12]

Thus far we have considered only the behavior of those

[12] One hardly needs to point out how important this behavior is in regard to the question of species formation from hybrids. (Footnote added later.)

pairs of traits in which one member dominates. The case of two or more differentiating traits also was discussed *theoretically* and tested *experimentally* by Mendel. It was shown that all possible combinations occur as frequently as they are expected on the basis of the laws of probability, assuming that their production is due to chance. "This also shows at the same time that *the behavior of each pair of differing traits in a hybrid association is independent of all other differences in the two parental plants.*" (Mendel)[13]

In the case of *two* pairs of traits, *nine* different classes of individuals may occur. However, only *four groups* may be distinguished *externally*; the numbers of individuals in the classes must occur in a ratio of 9:3:3:1. Among 1000 individuals, 562.5, 187.5, 187.5, and 62.5 respectively will be grouped together. In a suitable experiment Mendel did obtain the numbers 315, 101, 108, and 32 respectively, which on the basis of 1000 are as 566.6, 181.6, 194.2, and 57.6. This is a good approximation to the ratio. With hybrids of maize I have obtained the same result, in one case, for instance, the numbers 308, 104, 96, and 37 or, calculated on the basis of 1000, 565, 191, 176, and 68.

Mendel concludes "*that pea hybrids form germinal and pollen cells which, in their composition, correspond in equal numbers to all the constant*[14] *forms resulting from the combination of traits united through fertilization.*" We may say in the terms used in this paper: *In the hybrid, reproductive cells are produced in which the anlagen for the individual parental characteristics are contained in all possible combinations, but both anlagen for the same pair of traits are never combined. Each combination occurs with approximately the same frequency.* If the parental strains differ in only *one* pair of traits (2 traits: A, *a*) the hybrid will form only *two types*

[13] There are again exceptions to this rule; strains with linked traits exist. (Footnote added later.)

[14] Mendel calls a type constant, if it no longer contains the two different anlagen of a pair of traits.

of reproductive nuclei (A, *a*) which are like those of the parents. Each type is 50 percent of the total. If the parents differ in *two* pairs of traits (4 traits: A, *a*; B, *b*) *four types* of reproductive nuclei will be formed (*AB*, *Ab*, *aB*, *ab*), and 25 percent of the total will be of each type. If the parents differ in *three* pairs of traits (6 traits: A, *a*; B, *b*; C, *c*) *eight types* of reproductive nuclei will be formed (*ABC*, *ABc*, *AbC*, *Abc*, *aBC*, *aBc*, *abC*, *abc*), and 12.5 percent of the total will be of each type.[15]

This I call Mendel's *Law*. It includes the "loi de disjonction" of de Vries, also. Everything else may be derived from this law.

At present, however, this law is applicable only to a certain number of cases, i.e., those where one member of a pair of traits dominates,[16] and probably only to hybrids between varieties. It seems impossible that all pairs of traits of all hybrids should behave according to this law. Some hybrids of peas bear this out.

In the *first* generation of the combination of the "green, late [variety] Erfurter Folgererbse" with an almost colorless seed coat, and the "[variety] Kneifelerbse with purple-violet pods," or the "[variety] Pahlerbse with purple pods" both having a solid-color, orange-red seed coat, which turns brown on aging, the seed coats within the same pod were sometimes colorless, sometimes intensely red, *but usually more or less tinted with orange*, and also speckled to a variable degree with purplish-black spots. Thus, in addition to a *dilution* of one of the traits, an (apparently) *new* trait had originated. In the second generation, however, the seeds which show the two extremes of coloration, i.e., those with orange red, and those with almost colorless seed coats, will again produce

[15] If the pollen grains of the two parental strains differ externally, one may, if Mendel's Law holds, expect the hybrid to form two externally different types of pollen grains. That this is true was first observed by Focke.

[16] See footnote 6, p. 121.

the extreme types and all of the transitions between them. The speckling was sometimes unchanged, sometimes not present at all or very slight and sometimes somewhat increased. Size and shape of the seed and texture of the seed surface behaved in a similar way.

I will discuss these points at a later time.

Tübingen, 22 April, 1900.

POSTSCRIPT (ADDED IN PROOF)

In the meantime de Vries has published in these Proceedings (No. 3 of this year) some more details concerning his experiments. There he refers to Mendel's investigations, which *were not even mentioned in the "Comptes rendus."* I must emphasize again:

1. *that in many pairs of traits there is no dominating member* (p. 121),
2. *that Mendel's Law of segregation cannot be applied universally* (p. 130).

Tübingen, 16 May, 1900.

Part Four 1924-1925

Letters from H. de Vries and C. Correns to H. F. Roberts

In preparation for his book, *Plant Hybridization before Mendel* (1929), H. F. Roberts requested information from Hugo de Vries, Carl Correns, and Erich Tschermak on how they rediscovered Mendel's work. De Vries' reply in its original English, and translations of the German communications of Correns and Tschermak were published in his book. We are here reprinting the letters of de Vries and Correns: concerning Tschermak's contribution see p. x of the Foreword to this volume.

De Vries wrote in a letter dated December 18, 1924:

When preparing my book on the Mutation Theory, I worked on the basis of Darwin's Hypothesis of Pangenesis, and of the version of it proposed in my Intracellular Pangenesis. The main principle of Pangenesis is the conception of unit characters. This led on the one side to the theory of the origin of species by means of mutations, and on the other to the description of the phenomena of hybridization as recombinations of these units. In 1893, I crossed *Oenothera lamarckiana* with *O. lam. brevistylis*, and found their progeny to be uniform, and true to the specific parent in

1894, but splitting in the second generation 1895, giving 17–26 individuals with the recessive character (Mut. The. 11, p. 157). Many other species were tried with the same result, and dihybrid crosses showed the laws of chance to be valid for them also. After finishing most of these experiments, I happened to read L. H. Bailey's "Plant Breeding" of 1895. In the list of literature of this book, I found the first mention of Mendel's now celebrated paper, and accordingly looked it up and studied it.[1] Thereupon I published in March 1900 the results of my own investigations in the Comptes Rendus de l'Académie des Sciences, T. CXXX, p. 845, under the title of "Sur la loi de disjonction des hybrides," and shortly afterwards, in the same year, in the Berichte der deutschen botanischen Gesellschaft, T. XVIII, p. 83, (March 14, 1900). A full account of my experiments was given in the second volume of the German edition of my Mutation Theory, 1903.

Correns reported as follows (letter of January 23, 1925):
You ask further concerning the re-discovery of the Mendelian Laws. I cannot add much to what I have contributed in the Mendel issue of the "Naturwissenschaften." It will, in the meantime, certainly have reached your hands. The operation of a principle was soon found in the case of peas and maize. I was able accordingly soon to proceed systematically in the experiments, as the two genealogies for peas in my first contribution show. I did not come at first upon the explanation

[1] When de Vries wrote this letter in 1924 he had forgotten an earlier letter written by him personally to Bailey in 1905 or 1906. In this letter de Vries stated that he found the references to Mendel's papers in Bailey's article "Cross-Breeding and Hybridizing" of 1892. Mendel had indeed been cited in the bibliography which accompanied this article, in contrast to the first edition of Bailey's *Plant-Breeding* in which the 1892 article was reprinted without the bibliography. Only in 1902, in its second edition, did Bailey publish the bibliography. Obviously, de Vries' 1924 reference to Bailey's 1895 book was in error. See footnote in Roberts, p. 323, and Bailey's prefaces to the various editions of his book. (Eds.)

of the regular relationship (Gesetzmässigkeit), but I likewise, however, did not seek intensively after it. For I wished, for various reasons, to first finish an extensive book upon the sexual propagation of the foliose mosses, upon which I had worked for years. I then wished first to push intensively the elaboration of the investigations on xenias and hybrids, which had been carried on at the same time since 1894. The printing of the book lasted until in August 1899; then I was able to devote myself earnestly to the genetics researches. The date of the day upon which, in the autumn (October) of 1899, I found the explanation, I no longer know; I do not make note of such matters. I only know that it came to me at once "like a flash," as I lay toward morning awake in bed, and let the results again run through my head. Even as little do I know now the date upon which I read Mendel's memoir for the first time; it was at all events a few weeks later. I then first made ready for the press the contribution on xenias in maize. In it, it is already pointed out, that in crosses between maize races I had found very interesting but very complicated relationships. That other investigators also worked in the same direction I naturally did not know, otherwise I would have hastened more with the preparation of the publication.

On the morning of the 21st of April, 1900, I received a separate "Sur la loi de disjonction des hybrides," of De Vries, and by the evening of the 22nd of April, my contribution, "G. Mendels Regel über das Verhalten der Nachkommenschaft der Bastarde," was ready. I sent it to the German Botanical Society in Berlin, where it was received April 24, and was reported in the session of April 27. The issue in question of the "Berichte" appeared at the end of May, about the 25th. The contribution has been again printed in the volume in which the German Society for the Science of Heredity has recently collected my genetic works, insofar as they have not appeared independently anew.

For that matter, I do not lay too much weight upon the re-discovery itself. According to my opinion, it was important that the Mendelian laws should finally be known and verified. Whether it happened by their being independently found anew, or through the fact that someone first read the memoir of Mendel, and then repeated the experiments, is, however, at bottom, an indifferent matter *for science.* It was accordingly only a confirmation of what had been discovered more than 30 years before. And through all that in the meantime had been discovered and thought (I think above all of Weismann), the intellectual labor of finding out the laws anew for oneself was so lightened, that it stands far behind the work of Mendel. I myself should prefer, for my part, to lay more weight upon my later works, e.g., the *Bryonia* experiments.

In response to further inquiry, Professor Correns replied as follows (letter of January 30, 1925):

I did not discover the constant relationship (Gesetzmässigkeit) in *Pisum* alone but in *Zea* and *Pisum* simultaneously. In the publication, I placed *Pisum* in the foreground on Mendel's account, and out of didactic considerations. That I, however, also experimented with *Pisum* almost from the beginning, is explained from the way in which, as a matter of fact, I arrived at my genetic investigations. Originally I started out to solve the xenia question. To this end I wished to test experimentally all the assertions known in the literature. I began (1894) with *Phaseolus vulgaris nanus* (with which, however, cross-fertilization did not succeed at all for me), then with *Zea, Pisum, Lilium* and *Matthiola.* This is all related in my "Crosses between maize races, with particular reference to xenias," Bibliotheca Botanica, Heft 53 (1902), where the results are also mentioned; those upon *Matthiola* were also published previously in another place. For *Pisum* there are different pertinent assumptions. One only needs refer, for example, to Darwin, "The Variation of Animals and Plants under Domestication," and to Focke. Unfortunately Focke here, in the case of xenias, does not mention Mendel, otherwise I should have probably read his work immediately at the beginning of my investigations. After I had carried on cross-fertilizations with *Pisum* likewise on account of the xenia question (there exist, indeed, assumptions on the influence of the seed-coat), it was I suppose, quite natural to grow the crosses themselves, as I did not only with *Pisum*, but also with *Zea* and *Matthiola*, and finally also with *Lilium*. In this connection the advantages of *Pisum* naturally made themselves at once noticeable, especially the great convenience of the investigations, which, indeed, I could only carry on accessorily.

To one of my most fruitful objects of research, *Mirabilis jalapa*, I also first came indirectly, when I investigated (1900) the influence which the number of pollen grains used for pollination has upon the progeny. I had originally, indeed, not at all overlooked the matter of studying the behavior of the progeny in further generations, but had proceeded from other bases of inquiry.

Besides through Focke's book, I had been made cognizant of Mendel's investigations through my teacher Nägeli. And I believe also to remember that he told me of Mendel, but certainly only of the *Hieracium* investigations, in which alone he was permanently interested. Something of them was known to me also from the theoretical introduction to the first volume of the *Hieracium* monograph of Nägeli and Peter, and from Nägeli's introduction to the *Primula* monograph of E. Widmer. The memoir of Mendel on his *Hieracium* hybrids I first read, however, with that on the peas hybrids, in the autumn of 1899. Nägeli was, at the time when I became his pupil, already in ill health, read none of his colleagues' works any longer, and likewise no longer conducted his practicum any more. He interested me in the structure and growth of the vegetable cell membrane. When I began the genetic researches (1891), he was already dead. The above-cited references to Mendel, and indeed also the recollection of the verbal mention of Mendel, prompted me to ask Nägeli's family for possible letters received. His scientific correspondence was, however, not to be found at that time. It first came to light through an accident in 1904. The letters of Mendel were sent to me by the family, and were published by me; the remaining scientific correspondence the family then destroyed.

Part Five 1936-1966

Has Mendel's Work Been Rediscovered?

R. A. FISHER, M.A., SC.D., F.R.S.,
Galton Professor of Eugenics, University College, London

1. THE POLEMIC USE OF THE REDISCOVERY

The tale of Mendel's discovery of the laws of inheritance, and of the sensational rediscovery of his work thirty-four years after its publication and sixteen after Mendel's death, has become traditional in the teaching of biology. A careful scrutiny can but strengthen the truth in such a tradition, and may serve to free it from such accretions as prejudice or hasty judgment may have woven into the story. Few statements are so free from these errors as that which I quote from H. F. Roberts' valuable book *Plant Hybridisation before Mendel* (p. 286):[1]

The year 1900 marks the beginning of the modern period in the study of heredity. Despite the fact that there had been some

[This paper was originally published in Annals of Science 1 (1936): 115–137.]

[1] [Page numbers in *parentheses* refer to the work cited in the text. In Fisher's original publication several references are made to the English translation of Mendel's paper made in 1901 and reprinted in Bateson (1909). We have changed the page numbers to those of the new translation in this collection, and refer to them in *square brackets*, but have left the quotations from the 1901 translation unaltered. *Eds.*]

development of the idea that a living organism is an aggregation of characters in the form of units of some description, there had been no attempts to ascertain by experiment, how such supposed units might behave in the offspring of a cross. In the year above mentioned the papers of Gregor Mendel came to light, being quoted almost simultaneously in the scientific contributions of three European botanists, De Vries in Holland, Correns in Germany, and Von Tschermak in Austria. Of Mendel's two papers, the important one in this connection, entitled "Experiments in Plant Hybridization," was read at the meetings of the Natural History Society of Brünn in Bohemia (Czecho-Slovakia) at the sessions of February 8 and March 8, 1865. This paper had passed entirely unnoticed by the scientific circles of Europe, although it appeared in 1866 in the Transactions of the Society. From its publication until 1900, Mendel's paper appears to have been completely overlooked, except for the citations in Focke's *Pflanzenmischlinge*, and the single citation of Hoffmann, elsewhere referred to.

When the history of science is taken seriously the number of enquiries which such a story suggests is somewhat formidable. We want to know first: What did Mendel discover? How did he discover it? And what did he think he had discovered? Next, what was the relevance of his discoveries to the science of his time, and what was its reaction to them? In the case of Mendel these last questions must be duplicated, for we are concerned not only with the period following the reading of his principal paper in 1865, but with that following the widespread publicity it received in 1900. This will be considered first.

Seeing how often it is taken for granted that all clouds were cleared away at the rediscovery in 1900, it is singularly difficult to ascertain exactly how Mendel's experiments were conducted and, indeed, what experiments he carried out. We have, of course, his paper, principally devoted to garden peas, entitled "Versuche über Pflanzenhybriden," printed in the transactions of the Natural History Society of Brünn, in Bohemia, in 1866, and reprinted in 1910. In 1901 it was also

Ronald Aylmer Fisher, 1936.

twice reprinted, in *Flora* and in Ostwald's *Klassiker der exakten Wissenschaften* (No. 121). A valuable English translation, prepared for the Royal Horticultural Society, was published in 1901, and reprinted with modifications by Bateson on several occasions. I shall refer to its appearance in Bateson's book *Mendel's Principles of Heredity* (Cambridge, 1909).

It cannot be denied that Bateson's interest in the rediscovery was that of a zealous partisan. We must ascribe to him two elements in the legend which seem to have no other foundation: (1) The belief that Darwin's influence was responsible for the neglect of Mendel's work, and of all experimentation with similar aims; and (2) the belief that Mendel was hostile to Darwin's theories, and fancied that his work controverted them. On the first point we may note a paragraph from Bateson's preface (p. 2):

> While the experimental study of the species problem was in full activity the Darwinian writings appeared. Evolution, from being an unsupported hypothesis, was at length shown to be so plainly deducible from ordinary experiences that the reality of the process was no longer doubtful. With the triumph of the evolutionary idea curiosity as to the significance of specific differences was satisfied. The *Origin* was published in 1859. During the following decade, while the new views were on trial, the experimental breeders continued their work, but before 1870 the field was practically abandoned.

It should be noted that Bateson here identifies experimental breeding with the hybridization of species. He ignores the fact that Mendel's advance over his predecessors was due to crossing closely allied varieties, not different species, which, as Mendel actually recognized, would differ in a large number of different factors. It is a consequence of Darwin's doctrine that the nature of the hereditary differences between species can be elucidated by studying heredity in crosses within species. So far were the new evolutionary ideas from discouraging experimental breeding that Darwin, himself, apart from other work, devoted eleven years prior to 1876 to the great series of experiments of which his book, *The Effects of Cross- and Self-fertilisation in the Vegetable Kingdom,* is a report. Had his example been followed there would have been no such lull as succeeded his death. Like Mendel's experiments a few years earlier they seemed to lead to nothing more at the time. To-day, in the light of genetic analysis, we can go further towards appreciating their significance.

Bateson's eagerness to exploit Mendel's discovery in his feud with the theory of natural selection shows itself again in his misrepresentation of Mendel's own views. Although he was in fact not among those responsible for the rediscovery, his advocacy created so strong an impression that he is still sometimes so described. In the biographical notice which Bateson prefixes to his reprint of Mendel's papers he writes (p. 311):

> With the views of Darwin which were at that time coming into prominence Mendel did not find himself in full agreement, and he embarked on his experiments with peas, which as we know he continued for eight years.

The suggestion that Mendel was prompted by disagreement with Darwin's views to undertake his experiments is easily disproved. Mendel's experiments cannot have commenced later than 1857. Darwin's views on evolution were known only to a few friends prior to the papers which he communicated, jointly with Wallace, to the Linnean Society in 1858. That Mendel had heard of Darwin, as a geologist or an explorer, at the time his experiments with peas were commenced is, indeed, possible. More probably he knew nothing of Darwin's existence, and certainly nothing of the theory of Natural Selection, at this date. When, in 1865, Mendel reported his experiments, the situation had doubtless changed. Mendel now recognizes that the study of inheritance has a special importance in relation to evolutionary theory. He alludes to the subject, in his introductory remarks, in words which suggest not doubts, but rather a simple acceptance of the theory of evolution [p. 2].

> It requires indeed some courage to undertake a labour of such far-reaching extent; this appears, however, to be the only right way by which we can finally reach the solution of a question the importance of which cannot be overestimated in connection with the history of the evolution of organic forms.

In this paper the only other mention of evolution occurs in the concluding remarks, in which the results and opinions of Gärtner are discussed. It will be seen that Mendel expressly dissociates himself from Gärtner's opposition to evolution, pointing out on the other hand that Gärtner's own results are easily explained by the Mendelian theory of factors [p. 47].

> Gärtner by the results of these transformation experiments was led to oppose the opinion of those naturalists who dispute the stability of plant species and believe in a continuous evolution of vegetation. He perceives in the complete transformation of one species into another an indubitable proof that species are fixed within limits which they cannot change. Although this opinion cannot be unconditionally accepted we find on the other hand in Gärtner's experiments a noteworthy confirmation of that supposition regarding variability of cultivated plants which has already been expressed.

It is seen from these, the only two allusions to evolution in Mendel's paper, that he did not regard his work as a direct contribution to that subject. What he does claim for the laws of inheritance he established is that they make sense of many of the results of the hybridists, and that they form a necessary basis for the understanding of the evolutionary process. On this point he shows himself fully aware of the importance of what he had done. Had he considered that his results were in any degree antagonistic to the theory of selection it would have been easy for him to say this also.

2. SHOULD MENDEL BE TAKEN LITERALLY?

Bateson raised a point of great interest as to the conduct of Mendel's experiments in a footnote to a passage in the translation he used. After describing his first seven experiments Mendel opens his eighth (unnumbered) section with the words [p. 17], "In the experiments described above plants were used which differed only in one essential character (*wesentliches Merkmal*)."

Bateson notes:

> This statement of Mendel's in the light of present knowledge is open to some misconception. Though his work makes it evident that such varieties may exist, it is very unlikely that Mendel could have had seven pairs of varieties such that the members of each pair differed from each other in *only* one considerable character. The point is probably one of little theoretical or practical consequence, but a rather heavy stress is laid on the word *wesentlich*.

Most practical experimenters will feel the weight of this difficulty. Unless Mendel had known in advance of the separate inheritance of the characters he was studying he could scarcely have used seven such pairs of varieties. More probably, perhaps, he would have used fewer varieties, say four or five, and crossed these in all, six or ten, possible ways. In any case, we should expect that some or all of the crosses would have involved more than one contrasted pair of characters. Each progeny would then have segregated in more than one factor, and the question arises as to what Mendel did with these additional data. Two courses seem possible:

i) He might, for each cross, have chosen arbitrarily one factor, for which that particular cross was regarded as an experiment, and ignored segregation in other factors.

ii) He might have scored each progeny in all the factors segregating, assembled the data for each factor from the different crosses in which it was involved, and reported the results for each factor as a single experiment.

The first course seems incredibly wasteful of data. This objection is not so strong as it might seem, since it can be shown that Mendel left uncounted, or at least unpublished, far more material than appears in his paper. He evidently felt no anxiety lest his counts should be regarded as insufficient to prove his theory. But, apart from being wasteful, to have adopted this course would seem to imply as much foreknowledge of the outcome as if he had deliberately chosen unifactorial crosses. It would seem in any case an extremely arbitrary course to take.

The second course is in effect what most modern geneticists would do, unless they were discussing either the linkage or the interaction of more than one factor. Mendel nowhere gives summaries of the aggregate frequencies from different experiments, and this would be intelligible if the "experiments" reported in the paper were fictitious, being in reality themselves such summaries. Mendel's paper is, as has been frequently noted, a model in respect of the order and lucidity with which the successive relevant facts are presented, and such orderly presentation would be much facilitated had the author felt himself at liberty to ignore the particular crosses and years to which the plants contributing to any special result might belong. Mendel was an experienced and successful teacher, and might well have adopted a style of presentation suitable for the lecture-room without feeling under any obligation to complicate his story by unessential details. The style of didactic presentation, with its conventional simplifications, represents, as is well known, a tradition far more ancient among scientific writers than the more literal narratives in which experiments are now habitually presented. Models of the former would certainly be more readily accessible to Mendel than of the latter.

The great objection to the view suggested by Bateson's hint, that Mendel's "experiments" are fictitious, and that his paper is a didactic exposition embodying his accumulated data, lies in the words which Mendel himself used in introducing the successive steps of his account, e.g., at the beginning of the eighth section [p. 17], "The next task consisted in ascertaining . . . ," and the opening sentence of the ninth section [p. 23], "The results of the experiments previously described led to further experiments." It is true that the different experiments described are not numbered in a single series; those described in any one section are numbered afresh 1, 2, 3, . . . , so that these numbers were certainly assigned when the account was written; also we are never told in what year different plants were grown; yet, if Mendel is not to be taken literally, when he implies that one set of data was available when the next experiment was planned, he is taking, as *redacteur*, excessive and unnecessary liberties with the facts. Moreover, the style throughout suggests that he expects to be taken entirely literally; if his facts have suffered much manipulation the style of his report must be judged disingenuous. Consequently, unless real contradictions are encountered in reconstructing his experiments from his paper, regarded as a literal account, this view must be preferred to all alternatives, even though it implies that Mendel had a good understanding of the factorial system, and the frequency ratios which constitute his laws of inheritance, before he carried out the experiments reported in his first and chief paper. Such a reconstruction is attempted in the next section.

3. AN ATTEMPTED RECONSTRUCTION

A framework for dating the experiments is afforded by the statement [pp. 2–3], "This experiment was practically confined to a small plant group, and is now, after eight years, concluded in all essentials."

Mendel's paper was presented on the 8th of February, 1865; if he first grew his experimental peas in 1856[2] he

[2] [Fisher's reconstruction of the dates of Mendel's experiments led him to designate 1857 as the beginning and 1864 as the concluding year. Actually, the test period covered the years 1856–1863. (Mendel,

could then be reporting on eight seasons' work. His monastery had sent him for two years to the University at Vienna, where he had studied mathematics, physics, and biology. He returned and took up teaching duties in the Technical High School in 1853; he may then have undertaken work in the monastery garden for two years before starting his investigation of peas.

On this basis parts of the experiment can be definitely dated [p. 4]:

> In all thirty-four more or less distinct varieties of peas were obtained from several seedsmen and subjected to a two-years' trial . . . For fertilization twenty-one of these were selected and cultivated during the whole period of the experiments.

It was evidently in the second trial [experimental] year (1857) that the first cross-pollinations were made, namely, crosses for the two seed-characters *wrinkled* and *green*, and the two plant characters *white flowers* and *dwarf*. Of these the two first are said [pp. 15–16] to have shown segregation for six years, which must be 1858–63, the two named plant characters for five (1859–63), while the three other plant characters used by Mendel, *constricted pods, yellow pods*, and *terminal flowers*, for which only four segregating generations are mentioned, may have been first crossed a year later (1858).

In 1857 the recessiveness of the two seed characters must have appeared in the ripe seeds from the flowers cross-pollinated, for these would be round (or yellow) irrespective of the shape (or colour) of the self-fertilized seeds borne by the same plants. From the cross round by wrinkled sufficient seed was sown to raise 253 plants in 1858, while from the cross yellow by green 258 plants were raised. It is not im-

---

Letter to Nägeli, 18 April, 1867. See p. 160 of the present volume.) In this reprint of Fisher's paper all relevant years have been reduced by one. *Eds.*]

probable that about 250 plants heterozygous for each of the other two factors were also grown in 1858, but we are only told the numbers of plants raised from their seed in 1859, and these do not exceed what could have been bred from forty plants of each kind. In any case, ground for some 600 to 1000 cross-bred plants must have been needed in 1858, and it may be noted that in this year the number of self-fertilized lines was reduced from 38 to 22, releasing probably the ground occupied by sixteen rows.[3] The area of the experiments may well have been the same in the three years 1856, 1857, and 1858.

The heterozygous plants grown in 1858 from white-flowered parents, and those from dwarf parents, must have established the recessiveness of these characters, and so confirmed the fact of dominance in reciprocal crosses observed with the seed-characters in the previous year. In 1858 too, when the pods were ripe, seeds on plants heterozygous for *wrinkled* and *green* showed segregation in 3:1 ratios. For wrinkled seeds 253 plants gave 7324 seeds, an average of 29 to a plant. 5474 were round and 1850 wrinkled. The deviation from the expected 3:1 is less than its standard error of random sampling. For green seeds 258 plants gave 8023 seeds, an average of 31 to a plant. 6022 were yellow and 2001 green. The agreement with expectation is here even closer. Mendel does not test the significance of the deviation, but states the ratios as 2.96 : 1 and 3.01 : 1, without giving any probable error. The yield per plant seems low. Possibly only four or five pods on each plant were left to ripen, the remainder being consumed green; it is possible again that little room was allowed for each plant.

The discovery, or demonstration, whichever it may have

[3] [As noted by Bennett, Fisher seems to have mistaken the two years' testing period (1854–1855), which preceded the actual experiments, for the first two years of these experiments (1856–1857). Also, the reduction in the number of self-fertilized lines was from 34 to 22 (i.e., by 12) rather than from 38 to 22 (i.e., by 16). *Eds.*]

been, of the 3:1 ratio was evidently the critical point in Mendel's researches. The importance of the work was demonstrated, if not to Mendel himself, at least to his associates, and, in the following years, the area of the experimental site must have been greatly enlarged. Perhaps for the same reason, in this year also three new crosses were initiated, using the factors for *constricted pods*, *yellow pods*, and *terminal flowers*.

That Mendel was satisfied with the two approximate ratios so far obtained would be intelligible if, either previously or immediately upon reviewing the 1858 results, he had convinced himself as to their explanation, and framed the entire Mendelian theory of genetic factors and gametic segregation. His confidence and lack of scepticism shows itself in three distinct ways.

a) He has numerous opportunities in subsequent years of testing on a large scale whether or not the ratios really remained constant from year to year. If he made any such verification he does not record the data.

b) The test of significance of deviations from expectation in a binomial series had been familiar to mathematicians at least since the middle of the eighteenth century. Mendel's mathematical studies in Vienna may have given little attention to the theory of probability; but we know that he was engaged in other researches of a statistical character, in meteorology, and in connection with sun-spots, so that it is scarcely conceivable, had the matter caused him any anxiety, that he knew of no book or friend that would enable him to examine objectively whether or not the observed deviations from expectation conformed with the laws of chance. He goes so far as to give "by way of illustration" the classification of the seeds from "the first ten individuals" of each of these two series [p. 11]. In both cases the variations are no larger than the deviations to be expected, but Mendel does not say so. The average numbers of seeds from these two samples are above those for the whole series, being 44 against 29 in the first case and 48 against 31 in the second. Indeed, only three of the twenty plants give less than the average number for its experiment. Possibly some poor-yielding plants were rejected when the list was made up, in which case Mendel's statement, though it may be entirely honest, cannot be entirely literal. Possibly, again, the first ten plants had happened in each case to have been grown in more favourable conditions than the majority of the rest.[4]

Mendel also gives examples of extreme deviations in both directions from each series. These extreme cases, again, cannot be judged more extreme than would be expected among samples of about 250 plants, but Mendel gives no grounds for this opinion, and, indeed, does not express it.

c) The third point on which Mendel seems more incurious than we could imagine him being, were he not already satisfied, is in not comparing the outcome of reciprocal crosses. He alludes to the point at issue in a footnote to his concluding remarks [pp. 41–42].

In *Pisum* it is placed beyond doubt that for the formation of the new embryo a perfect union of the elements of both reproductive cells must take place. How could we otherwise explain that among the offspring of the hybrids both original types reappear in equal numbers and with all their peculiarities? If the influence of the egg-cell upon the pollen-cell were only external, if it

[4] I am obliged to Dr. J. Rasmussen, who has extensive experience of genetical work with *Pisum*, for the following explanation of Mendel's probable method of selection: "It is my impression that the classification was made throughout on dry plants in Winter. That is to say, that Mendel harvested his plants in Autumn, probably tied them up plot by plot, and for scoring loosened up the bunch of plants and picked out from it one plant after another. This is the method which first presents itself in work of this kind; it is also the method I am accustomed to use. The fact is that, working in this way, one will unconsciously choose the best plant first. This happens to me, whether I do the work myself or have other people picking out the plants from the bunch." In respect to the average yield Dr. Rasmussen also says: "About 30 good seeds per plant is, under Mendel's conditions (dry climate, early ripening, and attacks of *Bruchus pisi*) by no means a low number. It seems to me, indeed, rather a good one, and I feel convinced that Mendel classified all the seeds from these plants."

fulfilled the rôle of a nurse only, then the result of each artificial fertilization could be no other than that the developed hybrid should exactly resemble the pollen parent, or at any rate do so very closely. This the experiments have in nowise confirmed. An evident proof of the complete union of the contents of the two gametes is afforded by the experience gained on all sides that it is immaterial, as regards the form of the hybrid, which of the original species is the seed parent and which the pollen parent.

If, in 1858, any doubt as to the equivalence of the contributions of the two parents had entered Mendel's mind, he would surely have made a separate enumeration of the seeds borne by the two types of heterozygous plants derived from reciprocal pollinations. Their equivalence as regards dominance had been indicated in the previous year. Their equivalence in genic content Mendel seems early to have felt very sure of.

In 1930, as a result of a study of the development of Darwin's ideas, I pointed out that the modern genetical system, apart from such special features as dominance and linkage, could have been inferred by any abstract thinker in the middle of the nineteenth century if he were led to postulate that inheritance was particulate, that the germinal material was structural, and that the contributions of the two parents were equivalent. I had at that time no suspicion that Mendel had arrived at his discovery in this way. From an examination of Mendel's work it now appears not improbable that he did so and that his ready assumption of the equivalence of the gametes was a potent factor in leading him to his theory. In this way his experimental programme becomes intelligible as a carefully planned demonstration of his conclusions.

In 1859 the obstacles to the extension of his experimental programme had been overcome. In this year the two experiments with seed characters were completed by demonstrating that the 3:1 ratios observed in the previous year were genetically 1:2:1 ratios. In addition to an unknown number of wrinkled seeds, which came true for this character, 565 plants were raised from round seeds, of which 193 yielded round seeds only, while 372 behaved like their parents. Although at least a couple of pods from each of these 372 plants must have been allowed to ripen, the seed numbers are not reported and, perhaps, were not counted. In the second experiment some green seeds were sown, which duly gave green seeds only, while of 519 plants raised from yellow seeds 166 yielded yellow only and 353 were heterozygous. Again, no seed counts are reported from the 353 heterozygous plants. The ratios in both cases show deviations from the expected 2:1 ratio of less than their standard errors. This pair of experiments occupied the space of something more than 1084 plants. They were continued with smaller numbers for the next four years, but no further counts are given.

For the two plant characters *white flowers* and *dwarf*, which in this year (1859) first showed segregation, provision was made on a larger scale. Of 929 plants 224 bore white flowers, while of 1064 plants 277 were dwarfed. In both cases the deviation is less than the standard error of random sampling. In addition to making provision for over 3000 plants from the crosses made in 1857 Mendel must in this year have raised perhaps 250 heterozygous plants from each of the three crosses started in 1858. His cultures were therefore probably increased this year by about 3000 plants.

In 1860 provision was made for 1000 plants each for completing the experiments with the first two plant characters, these being families of 10 plants each from a hundred of the 1859 crop, chosen as showing the dominant characters, coloured flowers and tall stems respectively. The families from 36 plants had only coloured flowers, while those from 64 contained one or more white-flowered plants. The proportionate numbers among the 640 plants of these families were apparently not counted. Again, the families from 28 plants were exclusively tall, while 72 showed segregation of dwarfs. We are not told what was the frequency of dwarfs among these 720 plants. In neither case does the ratio depart significantly from the 2:1 ratio expected, although in the second case the deviation does exceed the standard deviation of random sampling.

In this year also the three crosses of plant characters started in 1858 required provision for nearly 1000 plants each. Of 1181 plants counted 299 had constricted pods, of 580 plants 152 had yellow pods, and of 858 plants 207 had terminal inflorescences. The deviation is below the standard in every case. Apart from progenies grown from recessive plants, these experiments account in all for 4619 plants. The total was thus probably greater than in the previous year, but the increase was not great.

So far as this, the first series of experiments, is concerned, there only remained in 1861 to provide for 3000 plants to establish the 2:1 ratios among the progenies of plants segregating for constricted pods, yellow pods, and terminal flowers. Out of a hundred parents tested there were respectively 29, 40, and 33 homozygous. Of these the first and third conform well with expectation. In the second case the observed frequencies, 40 homozygous to 60 heterozygous, show a relatively large, but not a significant, deviation. It is remarkable as the only case in the record in which Mendel was moved to verify a ratio by repeating the trial. A second series of a hundred progenies, presumably grown in 1862, gave 65:35, as near to expectation as could be desired. Although in 1860 only 580 plants had been available to display the 3:1 ratio for yellow pods, and in these two trials respectively 600 and 650 more must have appeared, they do not seem to have been counted, and are not reported in the paper.

In connection with these tests of homozygosity by examining ten offspring formed by self-fertilization, it is disconcerting to find that the proportion of plants misclassified by this test is not inappreciable. If each offspring has an independent probability, .75, of displaying the dominant character, the probability that all ten will do so is $.75^{10}$, or .0563. Consequently, between 5 and 6 percent of the heterozygous parents will be classified as homozygotes, and the expected ratio of segregating to nonsegregating families is not 2:1 but 1.8874 : 1.1126, or approximately 377.5 : 222.5 out of 600. Now among the 600 plants tested by Mendel 201 were classified as homozygous and 399 as heterozygous. Although these numbers agree extremely closely with his expectation of 200 : 400, yet, when allowance is made for the limited size of the test progenies, the deviation is one to be taken seriously. It seems extremely improbable that Mendel made any such allowance, or that the numbers he records as segregating are "corrected" values, rounded off to the nearest integer, obtained by dividing the numbers observed to segregate by .9437. We might suppose that sampling errors in this case caused a deviation in the right direction, and of almost exactly the right magnitude, to compensate for the error in theory. A deviation as fortunate as Mendel's is to be expected once in twenty-nine trials. Unfortunately the same thing occurs again with the trifactorial data.

These seven experiments of the first series require, as we have seen, a total of four or five thousand plants in the years 1859 and 1860. Apart from the continuation of heterozygous series they account for only 3000 in 1861 and for 1000 in 1862. The pollinations for his second series of experiments were, therefore, probably carried out in 1860. The large trifactorial experiment could not indeed have been finished had it started later, and, as the factor for white flowers first showed segregation in 1859, it is difficult to place it earlier. The bifactorial experiment took a year less, and might have been started in 1859, since the ripened seeds of 1858 had established the 3:1 ratios of the two factors. I shall suppose that both were initiated in 1860, and that the same is true of the important but smaller experiments devoted to determining the gametic ratios.

To 1861, then, are ascribed the fifteen doubly heterozygous plants of the bifactorial experiment, of which the 556 seeds displayed the first 9:3:3:1 ratio reported. All these were sown in 1862, even the thirty-two wrinkled-green seeds, which suggests that in this year space was abundant. (It was, indeed, in this same year that we have supposed Mendel to depart from his usual practice, and repeat the determination of a frequency ratio, at the expense of growing 1000 additional plants. Even with these additions the summary (Table VI) shows 1862 as less crowded than most of the other years.) The plants from these seeds, classified by the seeds they bore, exhibited independent segregation of the two factors. Mendel's classification of the 529 plants which came to maturity is shown in Table I.

TABLE I

Classification of Plants grown in the Bifactorial Experiment

	AA	A*a*	*aa*	Total
BB	38	60	28	126
Bb	65	138	68	271
bb	35	67	30	132
Total	138	265	126	529

The numbers are close to expectation at all points, but they are not very large. In relation to possible linkage, for example, they may be regarded as excluding, at the 5 percent level of significance, recombination fractions less than 44.9 per cent, which is not very strong negative evidence; yet on this point also Mendel evidently felt that further data would be superfluous, for he certainly could have obtained many more for the counting. The 138 plants, for example, recorded in the table above as being doubly heterozygous, doubtless bore over 4000 seeds segregating in the 9:3:3:1 ratio, and, even if the bulk of the crop were needed when green, at least ten seeds from each plant must have been allowed to ripen in order to classify the plant on which they grew.

The trifactorial experiment required 24 hybrid plants grown in 1861, which gave 639 offspring in 1862. In order to distinguish heterozygotes from homozygotes among the plants with coloured flowers progenies from at least 473 of these must have been grown. If, as in other cases, Mendel used a progeny of ten plants for such discrimination the experiment must have needed 4730 plants in 1863. Of this experiment Mendel says [p. 20], "Among all the experiments this demanded the most time and trouble," and the extent of the third filial generation explains this remark. It was evidently on the completion of this extensive work that Mendel felt that his researches were ripe for publication. It may have constituted the whole of his experimental work with peas in the last year before his paper was read.[5] Even so, probably this year saw more experimental plants than were grown in any previous year. Since the factor for coloured flowers used in this experiment obscures the cotyledon-colour of unopened seeds, not all of the vast number of seeds borne by these three generations was easily available to supplement the bifactorial and trifactorial data reported, yet even what was easily available must have been much more extensive than any data which Mendel published. Mendel's trifactorial classification of the 639 plants of the second generation is shown in Table II, which follows Mendel's notation, in which *a* stands for *wrinkled* seeds, *b* for *green* seeds, and *c for white flowers.*

In order to discriminate *CC* from *Cc* plants, progenies from these, which are seen to number 463 together, must have been grown on in 1863. In addition to abundant new unifactorial data the additional bifactorial data supplied by the experiments is seen to be large. 175 of the plants were heterozygous for both of the two seed characters, and, if 30 seeds from each had been classified, these would have given 5250 seeds, nearly ten times as many as the 556 reported

[5] [Actually two years before his paper was read. *Eds.*]

TABLE II

Classification of Plants grown in the Trifactorial Experiment

	CC				Cc			
	AA	Aa	aa	Total	AA	Aa	aa	Total
BB	8	14	8	30	22	38	25	85
Bb	15	49	19	83	45	78	36	159
bb	9	20	10	39	17	40	20	77
Total	32	83	37	152	84	156	81	321

	cc				Total			
	AA	Aa	aa	Total	AA	Aa	aa	Total
BB	14	18	10	42	44	70	43	157
Bb	18	48	24	90	78	175	79	332
bb	11	16	7	34	37	76	37	150
Total	43	82	41	166	159	321	159	639

from the bifactorial experiment. The classification of these plants as double heterozygotes must indeed have required that about half this number of seeds from each plant were examined. In the following year also nine-sixteenths of the progeny of 127 F_2 plants, or about 815 F_3 plants, must have borne seeds segregating in the 9:3:3:1 ratio, so that a further 24,000 seeds could have been so classified in 1863. Evidently, however, Mendel felt that the complete classification of 529 plants in the bifactorial experiment was sufficient; he does not even add, for the simultaneous segregation of *Aa* and *Bb*, the 639 plants completely classified in the trifactorial experiment, which suffice to raise the recombination fraction significantly higher than 46.56 per cent (from 44.9 per cent).

In the case of the 600 plants tested for homozygosity in

TABLE III

Comparison of Numbers reported with Uncorrected and Corrected Expectations

	Number of plants tested	Number of non-segregating progenies observed	Number expected		Deviation	
			Without correction	Corrected	Without correction	Corrected
1st group of experiments	600	201	200.0	222.5	+1.0	−21.5
Trifactorial experiment	473	152	157.7	175.4	−5.7	−23.4
Total	1073	353	357.7	397.9	−4.7	−44.9

the first group of experiments Mendel states his practice to have been to sow ten seeds from each self-fertilized plant. In the case of the 473 plants with coloured flowers from the trifactorial cross he does not restate his procedure. It was presumably the same as before. As before, however, it leads to the difficulty that between 5 and 6 per cent of heterozygous plants so tested would give only coloured progeny, so that the expected ratio of those showing segregation to those not showing it is really lower than 2:1, while Mendel's reported observations agree with the uncorrected theory.

The comparisons are shown in Table III. A total deviation of the magnitude observed, and in the right direction, is only to be expected once in 444 trials; there is therefore here a serious discrepancy.

If we could believe that Mendel changed his previous practice, and in 1862 went to the great labour of back-crossing the 473 doubtful plants, the data could be explained, for in such progenies misclassification would be only about one-fiftieth part as frequent as in progenies by self-fertilization. Equally, if we could suppose that larger progenies, say fifteen plants, were grown on this occasion, the greater part of the discrepancy would be removed. However, even using families of ten plants, the number required is more than Mendel had assigned to any previous experiment, and there is no reason for thinking that he ever grew so many as 7000 experimental plants in one year, apart from his routine tests.[6] Such explanations, moreover, could not explain the discrepancy observed in the first group of experiments, in which the procedure is specified, without the occurrence of a coincidence of considerable improbability.

An explanation of a different type is that the selection of plants for testing favoured the heterozygotes. In the first series of experiments the selection might have been made in the garden, or, if the whole crop was harvested, on the dry plants. In either case the larger plants might have been unconsciously preferred. It is also not impossible that, in some crosses at least, the heterozygotes may have been on the average larger than sister homozygotes. The difficulties to accepting such an explanation as complete are three. (*i*) In the trifactorial experiment there was no selection, for all plants grown must have been tested. The results here do not, however, differ in the postulated direction from those of the first series. On the contrary, they show an even larger discrepancy. (*ii*) It is improbable that the supposed compensating selection of heterozygotes should have been equally effective in the case of five different factors. (*iii*) The total compensation for all five factors (21.5 plants) must be supposed to be greater than would be needed (16.8 plants) if families of 11 had been grown, and less than would be needed (30.0) if 9 only had been grown, though nearly exactly right for the actual number 10 of F_3 plants in each progeny (22.5).

[6] The area available is given by Iltis as only 7 m by 35 m. Dr. Rasmussen estimates that he might have grown 4000–5000 plants in this area.

The possibility that the data for the trifactorial experiment do not represent objective counts, but are the product of some process of sophistication, is not incapable of being tested. Fictitious data can seldom survive a careful scrutiny, and, since most men underestimate the frequency of large deviations arising by chance, such data may be expected generally to agree more closely with expectation than genuine data would. The twenty-seven classes in the trifactorial experiment supply twenty-six degrees of freedom for the calculation of χ^2. The value obtained is 15.3224, decidedly less than its average value for genuine data, 26, though this value by itself might occur once in twenty genuine trials.

This total may be subdivided in various ways; one relevant subdivision is to separate the nine degrees of freedom created by the discrimination of homozygous and heterozygous plants with coloured flowers from the remaining seventeen degrees of freedom based on discriminations made presumably in the previous year. To the total the nine supply 6.3850, leaving only 8.9374 for the remaining seventeen. If anything, therefore, the subnormality in the deviations from expectation is more pronounced among the seventeen degrees of freedom than among the nine. If there has been sophistication there is no reason to think that it was confined to the final classification made in 1863.

To 1862 belong probably the bifactorial experiment and the five comparisons, each of four equal expected frequencies, supplied by the experiments on gametic ratios. The bifactorial experiment, having nine classes, supplies eight degrees of freedom for comparison, and gives a χ^2 of only 2.8110—almost as low as the 95 per cent point. The fifteen degrees of freedom of gametic ratios supply only 3.6730, should be included the verified 2:1 ratio for yellow pods, which is beyond the 99 per cent point. In the same year also giving 0.125 for one degree of freedom.

Putting together the comparisons available for 1862 we have:—

TABLE IV
Measure of Deviation Expected and Observed in 1862

	Expectation	χ^2 observed
Trifactorial experiment	17	8.9374
Bifactorial experiment	8	2.8110
Gametic ratios	15	3.6730
Repeated 2:1 test	1	0.1250
Total	41	15.5464

The discrepancy is strongly significant, and so low a value could scarcely occur by chance once in 2000 trials. There can be no doubt that the data from the later years of the experiment have been biased strongly in the direction of agreement with expectation.

One natural cause of bias of this kind is the tendency to give the theory the benefit of doubt when objects such as seeds, which may be deformed or discoloured by a variety of causes, are being classified. Such an explanation, however, gives no assistance in the case of the tests of gametic ratios and of other tests based on the classification of whole plants. For completeness it may be as well to give in a single table the χ^2 values for all the experiments recorded.

The bias seems to pervade the whole of the data, apart, possibly, from the illustrations of plant variation. Even the 14 degrees of freedom available before 1862 give only 7.1872, a value which would be exceeded about 12 times in 13 trials.

What I have inferred respecting the extent of Mendel's cultures is summarized by years in Table VI. I have arbitrarily allowed sixty plants for each of the stock lines and fifty for each segregating line which was continued with smaller numbers after the completion of the main experiments. I have included also in 1861 and 1862 the two small experiments devoted to the demonstration of gametic ratios. Some of the totals for years may be correct to the nearest

TABLE V

Deviations Expected and Observed in all Experiments

		Expectation		χ^2		Probability of exceeding deviations observed
3:1 ratios	Seed characters	2		0.2779		
	Plant characters	5	7	1.8610	2.1389	.95
2:1 ratios	Seed characters	2		0.5983		
	Plant characters	6	8	4.5750	5.1733	.74
Bifactorial experiment			8		2.8110	.94
Gametic ratios			15		3.6730	.9987
Trifactorial experiment			26		15.3224	.95
Total			64		29.1186	.99987
Illustrations of plant variation			20		12.4870	.90
Total			84		41.6056	.99993

TABLE VI

Approximate Numbers of Plants grown in Different Years[7]

	1856	1857	1858	1859	1860	1861	1862	1863
Stock lines	2280	2280	1320	1320	1320	1320	1320	1320
1st group	—	—	1011	3927	4719	3200	1350	350
2nd group	—	—	—	—	—	65	1719	4730
Total	2280	2280	2331	5247	6039	4585	4389	6400

[7] [Following Bennett, the total of 2280 plants assigned to 1856 and 1857 should be reduced in the ratio 34:38, thus yielding 2040 plants (cf. footnote 2, p. 149). *Eds.*]

hundred, but I do not expect all to be so. I feel justified in concluding only that the experiment was greatly enlarged after the first three years and that, with only ten plants to a family, the year 1863 was probably the fullest of all.

4. THE NATURE OF MENDEL'S DISCOVERY

The reconstruction has been undertaken in order to test the plausibility of the view that Mendel's statements as to the course and procedure of his experimentation are to be taken as an entirely literal account, or whether, on the other hand, there is evidence that data have been assembled from various sources, or the same data rediscussed from different standpoints in different sections of his account. There can, I believe, now be no doubt whatever that his report is to be taken entirely literally, and that his experiments were carried out in just the way and much in the order that they are recounted. The detailed reconstruction of his programme on this assumption leads to no discrepancy whatever. A serious and almost inexplicable discrepancy has, however, appeared, in that in one series of results the numbers observed agree excellently with the two to one ratio, which Mendel himself expected, but differ significantly from what should have been expected had his theory been corrected to allow for the small size of his test progenies. To suppose that Mendel recognized this theoretical complication, and adjusted the frequencies supposedly observed to allow for it, would be to contravene the weight of the evidence supplied in detail by his paper as a whole. Although no explanation can be expected to be satisfactory, it remains a possibility among others that Mendel was deceived by some assistant who knew too well what was expected. This possibility is supported by independent evidence that the data of most, if not all, of the experiments have been falsified so as to agree closely with Mendel's expectations.

The importance of the conclusion, if it is well established, that Mendel's statements are to be taken literally, lies in the inferences which flow from this view. *First,* that prior to the reported experiments Mendel was sufficiently aware of the independent inheritance of seven factors in peas to have chosen seven pairs of varieties, each pair differing only in a single factor. If it be thought that out of thirty-eight varieties he could not by deliberate choice have found the material for seven such crosses, it should be remembered also that at this stage he was choosing not only the varieties but, perhaps, also the factors to use in his experiment, and that he may have known of other factors in peas in addition to those with which his experiments are concerned, which, however, could not have been introduced without bringing in an undesirable complication.[8] *Next,* it appears that Mendel regarded the numerical frequency ratios, in which the laws of inheritance expressed themselves, simply as a ready method of demonstrating the truth of his factorial system, and that he was never much concerned to demonstrate either their exactitude or their consistency. It may be that the seed counts of 1858 were a revelation to him of the precision with which his system worked, and could be demonstrated; they may also possibly have given him an exaggerated impression of the precision with which the theoretical ratios should be verified, but from that moment it is clear, from the form his experiments took, that he knew very surely what to expect, and designed them as a demonstration for others rather than for his own enlightenment. That the hereditary contribution of the two parents might be unequal he did not seriously consider, although his first experiments provided splendid evidence on this important question, which it does not occur to

[8] It is particularly gratifying that this conclusion is supported by Dr. Rasmussen, basing his opinion upon existing types of garden peas, and on the development of these types since Mendel's time. He writes: "From the most probable assortment of varieties available to Mendel there would be no difficulty whatever in making unifactorial crosses in all characters. Indeed, the assortment at hand seems to have been much better fitted for such crosses than for other combinations."

him to present. It seems also not to have occurred to him that the inheritance of different factors might not be wholly independent. He asserts independence for all his factors, but gives evidence for only three of them, and for these much less than he might have given. A feature such as linkage would have been a complication extraneous to his theory, as he conceived it, which he would only have taken seriously had the observations forced it under his notice.

The theoretical consequences of his system he had thought out thoroughly, and in this respect his thought is considerably in advance of that of the first generation of geneticists which followed his rediscovery. He pointed out that n factors would give rise to 3^n different genotypes, of which 2^n would be capable of breeding true. He realized that even in intraspecific crosses n would be sufficiently great for these to be very large numbers, and that even more factors must be involved when crosses are made between different species, when minor in addition to major differences are considered. This understanding of the consequences of the factorial system contrasts sharply with many of the speculations of the earlier geneticists, such as that new species might be formed by the mutation of a single factor, or that the mimetic groups, found among butterflies and other insects, might be explained by the paucity of the genetic factors controlling the pattern and coloration of the wings. In these respects it has taken nearly a generation to rediscover Mendel's point of view.

Mendel seems also to have realized that the factorial system resolved one of the chief difficulties felt and discussed by Darwin, namely that, if the wide variation observable in cultivated plants were caused by the changed conditions and increased nourishment experienced on being brought into cultivation, then this cause of variation must continue to act, as Darwin had written, "for an improbably long time," since anciently cultivated species are not less but rather more variable than others. With segregating, heritable factors, on the other hand, the variability is easily explained by the preservation in culture of variants which, apart from man, would have been eliminated by natural selection. This, indeed, seems to have been Mendel's view [p. 37].

> It will be willingly granted that by cultivation the origination of new varieties is favoured, and that by men's labour many varieties are acquired which, under natural conditions, would be lost; but nothing justifies the assumption that the tendency to the formation of varieties is so extraordinarily increased that the species speedily lose all stability, and their offspring diverge into an endless series of extremely variable forms. Were the change in the conditions the sole cause of variability we might expect that those cultivated plants which are grown for centuries under almost identical conditions would again attain constancy. That, as is well known, is not the case, . . .

The reflection of Darwin's thought is unmistakable, and Mendel's comment is extremely pertinent, though it seems to have been overlooked. He may at this time have read the *Origin*, but the point under discussion may equally have reached his notice at second hand.

5. THE CONTEMPORARY REACTION TO MENDEL'S WORK

The peculiarities of Mendel's work, to which attention has been called in the previous sections, seem to contribute nothing towards explaining why his paper was so generally overlooked. The journal in which it was published was not a very obscure one, and seems to have been widely distributed. In London, according to Bateson, it was received by the Royal Society and by the Linnean Society. The paper itself is not obscure or difficult to understand; on the contrary, the new ideas are explained most simply, and amply illustrated by the experimental results. In view of the parallel failure of the biological world to appreciate and follow up Darwin's experiments, it is difficult to suppose that, had Mendel's paper been more widely read, there would have been many mentally prepared to appreciate its significance. Some there certainly were; and, had the new facts and methods come to the knowledge of Francis Galton, the experimental analysis of heredity might well have been established twenty-five years earlier than it was in fact; but minds equally receptive were certainly rare.

Among German biologists the one with whom Mendel is known to have corresponded is von Nägeli.[9] From his writings it is apparent either that Mendel's researches made no impression on his mind or that he was anxious to warn students against paying attention to them. In a paper published December 15, 1865, only ten months after the delivery of Mendel's paper on peas, and before its appearance in print, he seems to reprove observers who venture to think for themselves and to plan their own experiments instead of using the results of Gärtner and Kölreuter (p. 190):

> The knowledge of hybridization would in recent times have made more progress, if many observers, instead of beginning anew, had made use of the results of the two first-named German investigators, who applied the labour of their lives to the solution of this problem.

In the beginning of his paper Mendel had, with modest confidence, contrasted his method of procedure with that of these two distinguished predecessors. In his final discussion, also, he reinterprets the results of Gärtner in terms of the factorial system, showing that Gärtner's observations agreed with Mendel's theory, while dissenting from Gärtner's opinion that they were opposed to the theory of evolution.

In spite of his correspondence von Nägeli does not refer

[9] [This and the following paragraphs were written under the impression that Nägeli was aware of Mendel and his work when he published his paper of December 15, 1865. At that time, however, as pointed out by Fisher himself, Mendel's communication to the Brünn Society had not yet appeared in print and Mendel's opening letter in his correspondence with Nägeli was not to be written until more than two weeks later. It is most likely that Nägeli's comments were made without any knowledge of Mendel's existence. *Eds.*]

to Mendel's recent paper, and the following passage seems designed positively to ignore it (p. 231):

> Variability of the hybrids, that is to say, the diversity of forms which belong to the same generation, and their behaviour on propagation once or many times by self-fertilization, form two points in the study of hybridization which are still least ascertained, and which appear to be the least subject to strict rules.

Mendel had claimed to have established precisely such strict rules. Another passage in the same paper seems designed directly to contradict Mendel's claims as to the dominance and independence of genetic factors (p. 222):

> The characters of the parental forms are, as a rule, so transmitted that, in each individual hybrid both influences make themselves felt. It is not that one character is transmitted, as it were, unchanged from the one parent, a second unchanged from the other; but there occurs an interpenetration of the paternal and the maternal character, and a union between their characters.

It is difficult to suppose that these remarks were not intended to discourage Mendel personally, without drawing attention to his researches.

No such dishonourable intention can be ascribed to W. O. Focke, who, in his *Pflanzenmischlinge,* makes no less than fifteen references to Mendel. As in the case of other voluminous compilers, most of these references, though doubtless relevant to the different topics Focke had in mind, ignore the point of Mendel's work. The nearest Focke comes to giving any idea of what Mendel had done is found in the following sentence. This may stand as a good example of the limitations of even the best intentioned compilers of comprehensive treatises (p. [103]):

> Mendel's numerous crossings gave results which were quite similar to those of Knight, but Mendel believed that he found constant numerical relationships between the types of the crosses.

The fatigued tone of the opening remark would scarcely arouse the curiosity of any reader, and in all he has to say Focke's vagueness and caution have eliminated every point of scientific interest. Could any reader guess that the "constant numerical relationships" were the universal and concrete ratios of 1:1 and 3:1, or even that Focke was speaking of the frequency ratios of a limited number of recognizable genotypes?

It is not an accident that Focke was vague. In this case, as perhaps in others, he had not troubled to understand the work he was summarizing. Mendel's discovery of dominance and the great use he had made of seed characters had escaped him altogether. His comment continues:

In general, the seeds produced through a hybrid pollination preserve also, with peas, exactly the colour which belongs to the mother plant, even when from these seeds themselves plants proceed, which entirely resemble the father plant, and which then also bring forth the seeds of the latter.

H. F. Roberts makes an instructive comment on Focke's book:

A careful study of Focke's report brings into interesting relief the reason for his having failed to appraise the Mendel paper at its present value. In the first place, Focke was especially interested in the works of those who produced more extended contributions. The work of Kölreuter, Gärtner, Wichura and Wiegmann, whose works were much more voluminous in the field which they occupied, receive appropriate consideration, as do also Naudin's and Godron's prize contributions; but Mendel's paper evidently appeared to Focke simply in the guise of one of the numerous, apparently similar, contributions to the knowledge of the results of crossing within some single group . . . It was supposedly not at all conceivable that the laws of hybrid breeding could be compassed within a series of experiments upon a single plant.

Roberts ends his comment on a note of appreciation:

> The details of his (Focke's) data are laborious, exact, well-classified and scientifically arranged, comprising 79 families of dicotyledons, 13 families of monocotyledons, 2 families of gymnosperms, 2 of pteridophytes, one of the musci and one of the algæ.

It is very well to be reminded that the high qualities catalogued in the sentence last quoted are yet compatible with the learned author having overlooked, in his chosen field, experimental researches conclusive in their results, faultlessly lucid in presentation, and vital to the understanding not of one problem of current interest, but of many.

The peculiar incident in the history of biological thought which it has been the purpose of this study to elucidate, is not without at least one moral—namely, that there is no substitute for a careful, or even meticulous, examination of all original papers purporting to establish new facts. Mendel's contemporaries may be blamed for failing to recognize his discovery, perhaps through resting too great a confidence on comprehensive compilations. It is equally clear, however, that since 1900, in spite of the immense publicity it has received, his work has not often been examined with sufficient care to prevent its many extraordinary features being overlooked, and the opinions of its author being misrepresented. Each generation, perhaps, found in Mendel's paper only what it expected to find; in the first period a repetition of the hybridization results commonly reported, in the second a discovery in inheritance supposedly difficult to reconcile with continuous evolution. Each generation, therefore, ignored what did not confirm its own expectations. Only a succession of publications, the progressive building up of a *corpus* of scientific work, and the continuous iteration of all new opinions seem sufficient to bring a new discovery into general recognition.

BIBLIOGRAPHY[10]

W. BATESON

1909 Mendel's Principles of Heredity. Cambridge University Press.

C. DARWIN

1859 The Origin of Species. London, John Murray.

C. DARWIN

1876 The Effects of Cross- and Self-Fertilisation in the Vegetable Kingdom. London, John Murray.

R. A. FISHER

1930 The Genetical Theory of Natural Selection. Oxford, Clarendon Press.

W. O. FOCKE

1881 Die Pflanzen-Mischlinge. Berlin, Borntraeger.

H. ILTIS

1924 Gregor Johann Mendel Leben, Werk, und Wirkung. Berlin, Julius Springer.

G. MENDEL

1866 Versuche über Pflanzen-Hybriden. Verhandlungen des naturforschenden Vereines in Brünn, 4 (1865) (Abh.): 3–47.

C. NÄGELI

1866 Die Bastardbildung im Pflanzenreiche. Botanische Mittheilungen 2: 187–235.

H. F. ROBERTS

1929 Plant Hybridization before Mendel. Princeton, N.J., Princeton University Press.

[10] [A number of errors in the original bibliography have been corrected. *Eds.*]

Mendel's Ratios

SEWALL WRIGHT
Department of Genetics, University of Wisconsin, Madison, Wisconsin

The excessive goodness of fit of Mendel's ratios is certainly one of the most disconcerting items that a historian of genetics has to deal with. I have repeated the χ^2 test from an independent tabulation and come out with substantially the same result as Fisher.

There is no question that the data fit the ratios much more closely than can be expected from accidents of sampling. Fisher seems to assume that this necessarily implies deliberate falsification of the data and suggests that this was by an assistant.

I do not think that Fisher allows enough for the cumulative effect on χ^2 of a slight subconscious tendency to favor the expected result in making tallies. Mendel was the first to count segregants at all. It is rather too much to expect that he would be aware of the precautions now known to be necessary for completely objective data. Anyone who doubts the difficulty in making repeatable counts should read chapter 5 in Pearl's *Introduction to Medical Biometry*. He reports an experiment in which 15 trained observers obtained extraordinary differences in sorting and counting the same 532 kernels of corn. Checking of counts that one does not like, but not of others, can lead to systematic bias toward agreement. I doubt whether there are many geneticists even now whose data, if extensive, would stand up wholly satisfactorily under the χ^2 test.

The most serious evidence for fraud by Mendel, presented by Fisher, is the very close agreement to a ratio of 1AA:2A*a* in the tests of F_2 for plant traits for one-factor crosses and for that of seed coat in his three-factor cross, made by looking for segregation in groups of 10 progeny from dominant F_2's. Fisher points out that $(3/4)^{10}(= .0563)$ of the A*a* group include no recessives and would be classified as if AA, making the expected ratio .371AA : .629A*a*.

Mendel, however, reported the ratios 43:2 and 14:15 as the extreme F_2's on F_1 plants; round × angular, and 32:1 and 20:19 as the extreme F_2's in F_1 yellow × green. These are so difficult to account for among 511 F_1 plants as to suggest some complication, but in any case, they would hardly have been reported by one bent on fraud. They cannot, of course, be included in the χ^2 test since they are selected extreme cases.

Having observed such ratios as 43:2 and 32:1, Mendel could hardly have been unaware of the likelihood that no recessives would appear in some groups of 10 progeny from A*a* and was certainly capable of estimating that the chance that they would appear was about one in eighteen. His use of the obviously rather inadequate number 10 must have been a compromise, forced by the desire to test 100 F_2's in limited space. It would seem likely that his statement that the groups contained 10 plants is not to be taken wholly literally. He knew that there could be losses and may well have planted somewhat more than 10 in each case to be sure of having at least 10 to examine for segregants. If the average was 12, the probability of misclassifying A*a* falls from .056 to .031.

More important, perhaps, is the likelihood that he could often distinguish a segregating group (3.3AA : 6.7A*a*) from a nonsegregating one (10AA) even in the absence of any recessives. From his description of the heterozygotes for grey and white seed coat there would seem little doubt that in this case at least the occurrence of segregation of AA and *Aa* would be obvious in a group of 10 in the absence of recessives. I suspect that he used seed coat with the two real seed traits in his three-factor cross for this reason.

With respect to the plant traits in general he never insists on complete identity of AA and *Aa* but merely on the unreliability of classifying single plants. The classification of a progeny of 10 or more plants as from AA or *Aa* would be another matter. In short, I doubt whether Mendel in his very succinct account has given in full detail his criteria for distinguishing progenies from AA and *Aa*.

In the 91 percent of the data concerned with one-factor ratios and their testing, the amount of bias toward expectation is slight and on a rough estimate would need less than 2 such misentries in 1000 to reduce χ^2 below expectation as much as observed. The more complicated two- and three-factor tests and especially the tests of gametic ratios involve more serious bias, amounting to some 3 percent possibly directed misentries in the last case and about 1.7 percent collectively. In tallying those small experiments, Mendel must have been conscious of how the rows were running, especially where the expected ratio was 1:1:1:1 in the tests of gametic ratios (probability of worse fit is about .002 in contrast to .05 in the worst of the other cases) and I am afraid that it must be concluded that he made occasional subconscious errors in favor of expectation, especially in this case. Taking everything into account, I am confident, however, that there was no deliberate effort at falsification.

Literature

BAILEY, L. H.

1892 Cross-Breeding and Hybridizing. Rural Library 1:3–44. Rural Publishing Co., New York.

1895 Plant-Breeding. Macmillan, New York. 293 pp.

BATESON, W.

1902 Mendel's Principles of Heredity. A Defence. Cambridge University Press. 212 pp.

1909 Mendel's Principles of Heredity. Cambridge University Press. 396 pp.

BENNETT, J. H., *ed.*

1965 Experiments in Plant Hybridisation: Gregor Mendel. Oliver & Boyd, Edinburgh. 95 pp.

CORRENS, C.

1900 G. Mendel's Regel über das Verhalten der Nachkommenschaft der Rassenbastarde. Berichte der deutschen botanischen Gesellschaft 18:158–168. English translation *in* The Birth of Genetics. Genetics 35, no. 5, pt. 2 (1950):33–41.

1905 Gregor Mendel's Briefe an Carl Nägeli, 1866–1873. Abhandlungen der mathematisch-physischen Klasse der königlich sächsischen Gesellschaft der Wissenschaften 29:189–265. English translation *in* The Birth of Genetics. Genetics 35, no. 5, pt. 2 (1950): 1–29.

EDITORS OF GENETICS

1950 The Birth of Genetics. Supplement to Genetics, Volume 35, no. 5, pt. 2. 47 pp.

FISHER, R. A.

1936 Has Mendel's work been rediscovered? Annals of Science 1:115–137.

FOCKE, W. O.

1881 Die Pflanzen-Mischlinge. Gebr. Borntraeger, Berlin. 569 pp.

GÄRTNER, C. F. *von*

1849 Versuche und Beobachtungen über die Bastarderzeugung im Pflanzenreich. Stuttgart. 791 pp.

JAKUBÍČEK, M., *and* J. KUBÍČEK

1965 Bibliographia Mendeliana. Universitní knihovna v Brné, Brno. 74 pp.

MENDEL, G.

1866 Versuche über Pflanzen-Hybriden. Verhandlungen des naturforschenden Vereines in Brünn 4 (1865): Abh., 3–47. Reprinted 1901 *in* Ostwald's Klassiker der exakten Wissenschaften (E. Tschermak, ed.) No. 121: 3–46, Engelmann, Leipzig; 1901 *in* Flora 89: 364–403; 1951 *in* Journal of Heredity 42: 3–47. Original manuscript reproduced *in* Gedda, L. (1956) Novant' anni delle Leggi Mendeliane: 5–99. Inst.

Gregorio Mendel, Rome. English translation first published in 1901 *in* Journal of the Royal Horticultural Society 26:1–32.

1869 Uber einige aus künstlicher Befruchtung gewonnenen Hieracium-Bastarde. Verhandlungen des naturforschenden Vereines in Brünn 8(1868): Abh., 26–31. Reprinted 1901 *in* Ostwald's Klassiker der exakten Wissenschaften (E. Tschermak, ed.) No. 121: 47–53. Engelmann, Leipzig. English translation first published *in* Bateson (1902).

NÄGELI, C.

1865 Die Bastardbildung im Pflanzenreiche. Sitzungsberichte der königlichen bayerischen Akademie der Wissenschaften zu München, mathematisch-physikalische Classe (Sitzgber. k.b. Akad. Wiss. Munich) 1865 (2):395–443. Also in C. Nägeli, Botanische Mittheilungen 2:187–235 (1866).

1866 Ueber die abgeleiteten Pflanzenbastarde. Sitzgber. k.b. Akad. Wiss. Munich 1866 (1):71–93. Also in Botanische Mittheilungen 2:237–259 (1866).

1866 Die Theorie der Bastardbildung. Stizgber. k.b. Akad. Wiss. Munich 1866 (1):93–127. Also in Botanische Mittheilungen 2:259–293 (1866).

1866. Ueber die Zwischenformen zwischen den Pflanzenarten. Sitzgber. k.b. Akad. Wiss. Munich 1866 (1): 190–221. Also in Botanische Mittheilungen 2:294–325 (1866).

1866 Aufzählung einiger Zwischenformen. Sitzgber. k.b. Akad. Wiss. Munich 1866 (1):222–235. Also in Botanische Mittheilungen 2:326–339 (1866).

ROBERTS, H. F.

1929 Plant Hybridization before Mendel. Princeton University Press. 374 pp.

STOMPS, T. J.

1954 On the rediscovery of Mendel's work by Hugo de Vries. Journal of Heredity 45:293–294.

TSCHERMAK, E.

1900 Über künstliche Kreuzung bei *Pisum sativum.* Berichte der deutschen botanischen Gesellschaft 18: 232–239.

1900 Über künstliche Kreuzung bei *Pisum sativum.* Zeitschrift für das landwirthschaftliche Versuchswesen in Oesterreich 3:465–555.

1900 Über künstliche Kreuzung bei *Pisum sativum.* Biologisches Zentralblatt 20:593–595. English translation *in* The Birth of Genetics. Genetics 35, no. 5, pt. 2 (1950):42–47.

DE VRIES, H.

1900 Das Spaltungsgesetz der Bastarde. Vorläufige Mitteilung. Berichte der deutschen botanischen Gesellschaft 18:83–90.

1900 Sur la loi de disjonction des hybrides. Comptes rendues de l'académie des sciences 130:845–847. English translation *in* The Birth of Genetics. Genetics 35, no. 5, pt. 2 (1950):30–32.

1900 Sur les unités des caractéres spécifiques. Revue générale de botanique 12:257–271.

WICHURA, M.

1865 Die Bastardbefruchtung im Pflanzenreich, erläutert an den Bastarden der Weiden. E. Morgenstern, Breslau. 95 pp.